Synthesis Lectures on Mechanical Engineering

This series publishes short books in mechanical engineering (ME), the engineering branch that combines engineering, physics and mathematics principles with materials science to design, analyze, manufacture, and maintain mechanical systems. It involves the production and usage of heat and mechanical power for the design, production and operation of machines and tools. This series publishes within all areas of ME and follows the ASME technical division categories.

Allan T. Kirkpatrick

Introduction to Refrigeration and Air Conditioning Systems

Theory and Applications

Second Edition

 Springer

Allan T. Kirkpatrick
Colorado State University
Fort Collins, CO, USA

ISSN 2573-3168   ISSN 2573-3176   (electronic)
Synthesis Lectures on Mechanical Engineering
ISBN 978-3-031-16778-2   ISBN 978-3-031-16776-8   (eBook)
https://doi.org/10.1007/978-3-031-16776-8

This Springer imprint is published by the registered company Springer Nature Switzerland AG
The registered company address is: Gewerbestrasse 11, 6330 Cham, Switzerland

Preface

The second edition builds on the foundation established by the previous first edition published in 2017. The first edition covered background information, description, and analysis of four major cooling system technologies—vapor compression cooling, evaporative cooling, absorption cooling, and gas cooling. The second edition has been expanded to include increased coverage of cooling system refrigerants, fluid mechanics, heat transfer, and building cooling loads. With increasing climate change due to the buildup of greenhouse gas emissions in the atmosphere, there has been a worldwide impetus to transition to cooling systems and refrigerants that have a low or even zero global warming potential. A short review of refrigeration and air conditioning is given in Chapter One. Chapter Two discusses the analysis and performance of vapor compression systems, while Chapter Three covers evaporative, absorption, and gas cooling systems. Chapter Four reviews the relevant fluid mechanics, and Chapter Five reviews the heat transfer fundamentals. The determination of building cooling loads is the subject of Chapter Six. Finally, Chapter Seven provides detailed information and analysis of the components—heat exchangers, pumps, and compressors, used to assemble a cooling system.

The text is written as a tutorial for engineering students and practicing engineers who want to become more familiar with the performance of refrigeration and air conditioning systems. The goals are to familiarize the reader with cooling technology nomenclature, and provide insight into how refrigeration and air conditioning systems can be modeled and analyzed. Emphasis is placed on constructing idealized thermodynamic cycles to represent actual physical situations in cooling systems. The book contains numerous practical examples to show how one can calculate the performance of cooling system components. By becoming familiar with the analyses presented in the examples, one can gain a feel for the representative values of the various thermal and mechanical parameters that characterize cooling systems.

Fort Collins, CO, USA   Allan T. Kirkpatrick
August 2022

Acknowledgements

Discussions with Colorado State Professors Daniel Wise and Bret Windom about refrigeration and air conditioning issues have been very helpful. Former CSU graduate students Charles Boardman, Michael Deru, Kevin Knappmiller, and Joel Neymark worked tirelessly with me on building energy use research problems.

Many thanks to the editorial staff at Springer Nature for their assistance with the second edition. Mr. Paul Petralia deserves special acknowledgment for his encouragement of this project. I would like to thank my wife Susan and my extended family, Anne, Matt, Maeve, Michael, Rob, Kristin, Thomson, Charlotte, and Theo for their unflagging support while the second edition was being written.

Finally, this edition is dedicated to my late father, Edward T. Kirkpatrick, who sparked my interest in mechanical engineering years ago.

Contents

About the Author

Dr. Allan T. Kirkpatrick is currently an Emeritus Professor in the Mechanical Engineering Department at Colorado State University. He has BS (1972) and Ph.D. (1981) degrees in Mechanical Engineering from the Massachusetts Institute of Technology. He is an internationally recognized authority in the applied thermal-fluid sciences and engineering education areas, and has published five books and over 110 publications in these areas. He has taught undergraduate and graduate courses in thermodynamics, building energy systems, heat transfer, and fluid mechanics, and has been an active member of ASME and ASHRAE. Dr. Kirkpatrick is a Fellow of the American Society of Mechanical Engineers, and has been recognized over his career with numerous national and international awards, most recently the ASME Ben Sparks Medal.

Introduction to Cooling Technologies

1

1.1 Introduction

This book provides background information, description, and analysis of four major cooling technologies—vapor compression cooling, evaporative cooling, absorption cooling, and gas cooling. Vapor compression systems are currently the primary technology used in most standard domestic, commercial, and industrial cooling applications, as they have both performance and economic advantages over the other competing cooling systems. However, there are many other applications in which evaporative cooling, absorption cooling, and gas cooling technologies are a preferred choice.

Cooling technologies are generally divided into air conditioning and refrigeration applications. Air conditioning technologies are defined as those that are used to maintain acceptable thermal comfort conditions for people and equipment in residential, commercial, and industrial buildings and spaces, typically in the neighborhood of 20–30 ° C. Refrigeration technologies are defined as those that are used to maintain temperatures near or below freezing (0° C) for safe storage of perishable items such as food and medicine, and operation of low-temperature laboratory equipment.

For the most part, the vapor compression and absorption technologies can be used for either air conditioning or refrigeration applications by simply changing the temperature of the refrigerant in the evaporator and the condenser. The evaporation and gas cooling technologies are primarily used for air conditioning applications.

The main focus of the text is on the application of the thermal sciences to refrigeration and air conditioning systems. The goals are to familiarize the reader with cooling technology nomenclature, and provide insight into how refrigeration and air conditioning systems can be modeled and analyzed. With the need to reduce greenhouse gas emissions, there has been increased attention paid to reducing the environmental impact of refrigeration and air conditioning systems. Cooling systems are inherently complex, as the second law

© The Author(s), under exclusive license to Springer Nature Switzerland AG 2023
A. T. Kirkpatrick, *Introduction to Refrigeration and Air Conditioning Systems*,
Synthesis Lectures on Mechanical Engineering,
https://doi.org/10.1007/978-3-031-16776-8_1

of thermodynamics does not allow thermal energy to be transferred directly from a lower temperature to a higher temperature, so the heat transfer is done indirectly through a thermodynamic cycle. Emphasis is placed on constructing idealized thermodynamic cycles to represent actual physical situations in cooling systems.

The text is written as a tutorial for engineering students and practicing engineers who want to become more familiar with air conditioning and refrigeration systems. The level of the text is at the advanced mechanical engineering student level. It assumes basic knowledge of thermodynamic properties, open system equations, and psychrometrics. A short review of these topics is given in Chapter One. Chapter Two discusses the analysis and performance of vapor compression systems, while Chapter Three covers evaporative, absorption, and gas cooling systems. Chapter Four reviews the relevant fluid mechanics, and Chapter Five reviews the heat transfer fundamentals. The determination of building cooling loads is the subject of Chapter Six. Finally, Chapter Seven provides detailed information and analysis of the components—heat exchangers, pumps, and compressors—used to assemble a cooling system.

The text contains numerous practical examples to show how one can calculate the performance of cooling systems. By becoming familiar with the analyses presented in the examples, one can gain a feel for the representative values of the various thermal and mechanical parameters that characterize cooling systems.

1.1.1 Cooling Technologies

Cooling technologies take advantage of both the sensible heat and the latent heat of a working fluid. Sensible heat transfer is determined by the property defined as specific heat as that represents the energy (4.18 kJ/kg-K for water) needed to increase or decrease an object's temperature without changing its thermodynamic phase. Latent heat transfer is determined by the property defined as the heat of vaporization or fusion (2230 kJ/kg for water) as that represents the energy needed for a phase change from a solid to a fluid phase or from a fluid to a gas phase. The melting of a block of ice will require heat transfer from another body producing a cooling effect on that body. Likewise, the evaporation of a fluid will require heat transfer from another body, also producing a cooling effect. Latent heat transfer is used directly in evaporative cooling systems and in the evaporator and condenser components of vapor compression systems.

A mechanical vapor compression cycle uses the evaporation of a refrigerant or working fluid to produce a net cooling effect. There are many choices for the refrigerant in vapor compression cycles. The major requirement is that the fluid thermal properties match the requirements of the given application. There is a narrow temperature range for vaporization and condensation, and the refrigerant should also have a relatively high heat of vaporization.

Other important considerations are global warming potential, flammability, toxicity, stability, cost, lubricant and materials compatibility, and environmental impact. With increasing climate change due to the buildup of greenhouse gas emissions in the atmosphere, there has been a worldwide impetus to transition to refrigerants that have a low or even zero global warming potential. We are on pace for about $4\,°C$ of warming by the end of the century unless we reduce our greenhouse gas and CO_2 emissions. As will be discussed later, there is no "perfect" refrigerant, and the choice for a particular application will necessarily involve compromises.

An absorption cycle uses external heating of the working fluid to produce a cooling effect. The system components are more complex, thus more expensive, in an absorption system relative to a vapor compression system, so absorption systems are used mainly in large industrial and commercial building applications and in situations where electricity is not available, such as vacation homes, mobile homes, and trailers. There are two working fluids in an absorption cycle, an absorbent and a refrigerant. The absorption cycle uses the solubility of the refrigerant gas in the absorbent liquid to reduce the pumping energy required to compress the refrigerant.

Evaporative cooling is used when low-humidity air is available, and can be cooled by the evaporation of water sprayed into the air stream. It is less expensive to install and maintain relative to a vapor compression system, and has lower power consumption, since no compressor is needed. The working fluid is water, not a halocarbon refrigerant. However, the temperature decrease of the air stream is smaller when compared to vapor compression cooling.

1.2 Brief History of Cooling Technologies

1.2.1 Refrigeration

Vapor compression systems were first commercialized in the early part of the twentieth century for refrigeration applications and then adopted for air conditioning applications. Before the introduction of vapor compression technology, ice was used as the main method of refrigeration. In winter months, blocks of ice were sawed from frozen riverbeds and stored in insulated warehouses. The ice was then periodically transported to commercial facilities and residences. The blocks were placed near the ceiling of the enclosure, and the resulting natural convection cold air currents provided cooling and transferred thermal energy away from the object or space to be cooled. Blocks of ice were also placed in insulated cabinets to maintain a safe temperature for the storage of food. Similar ice storage techniques were used in passenger trains to provide air conditioning for the occupants.

In the 1800s, numerous inventors and engineers attempted to develop vapor compression for refrigeration applications. Jacob Perkins (1766–1849), an American inventor, built the first working vapor compression refrigeration system in 1834. It used ether as the working fluid and was a closed-cycle that could operate continuously, however, it was not successfully commercialized for industrial use until the late 1800s. In 1851, James Harrison, a Scottish inventor, developed an ether-based vapor compression refrigeration system that was used in the 1880s to transport frozen meat in ships from Australia to England. Since the transport time to market was on the order of months, this refrigeration technology vastly increased the ability of meat-producing countries to participate in world trade.

Beginning in 1900, there was a gradual decline in commercial ice houses and a greater use of on-site vapor compression refrigeration. Industrial vapor compression refrigerators in the early 1900s used either ammonia (NH_3) or methyl chloride (CH_3Cl) as refrigerants. The toxicity of these working fluids restricted their use for residential refrigeration and motivated the development of alternative non-toxic refrigerants in the early 1920s. As a consequence, commercial refrigeration using vapor compression systems preceded domestic refrigeration by about 50 years.

In 1918, General Electric developed a refrigerator with the motor enclosed in the compressor case, which prevented leaks between the motor shaft and compressor. In 1923, air-cooled condensers for domestic refrigerators were introduced, replacing the water-cooled condenser which had to be connected to a wastewater line. The domestic refrigerators used sulfur dioxide (SO_2) as the working refrigerant, which has a low working pressure and is comparatively non-toxic and nonflammable. The domestic refrigerators were configured as split systems, with the air-cooled condenser located in a basement or outside.

After the successful development of non-toxic halocarbons, vapor compression systems began to be widely adopted for domestic refrigeration and air conditioning applications. With increased electrification of the United States, the percentage of residences in the US with electrically powered vapor compression refrigerators increased from 44% in 1940 to over 98% by 1960 (Gordon 2016).

The absorption cooling cycle was invented in 1858 by Ferdinand Carre (1824–1900), a French scientist. His cycle used water and ammonia to produce ice and preserve food. With the invention of ammonia absorption cooling, ice was manufactured year-round and used for refrigeration and air conditioning. In 1900, most ice plants used ammonia absorption technology to produce ice for both residential and commercial use. In 1925, gas-fired absorption refrigerators for residential use were commercialized and marketed by the Electrolux Corporation, and widely adopted in rural areas without electricity. Absorption systems are now primarily used in remote locations which have not been electrified.

1.2.2   Air Conditioning

Before the incorporation of air conditioning systems in buildings, most spaces were cooled using natural ventilation through windows, incorporation of heavy thermal mass in the building structure, and evaporative cooling. The first mechanical air conditioning system for buildings was developed by Willis Carrier (1876–1950), an American mechanical engineer in 1903. His system used a water spray to control the humidity and dew point temperature in a space, and filter out dust particles. The paper and textile industries were the earliest adopter of his water spray air conditioning system, followed by hospitals.

In 1923, the Carrier corporation commercialized a vapor compression cooling system using dichloroethene as the working fluid. Movie theaters and department stores were early adopters of this air conditioning system in order to obtain a business advantage over competitors. By 1930, most government, retail, and office buildings, as well as passenger trains, were air conditioned using vapor compression systems.

There has been a rapid penetration of air conditioning systems into residential buildings and vehicles over the last 50 years. For example, the percentage of residences in the US with central or room air conditioning systems has increased from about 37% in 1970 to 89% in 2010. Likewise, the percentage of vehicles with air conditioning has increased from about 20% in 1960 to 84% in 1983, Gordon (2016). The air conditioning technology employed for vehicles has been vapor compression. Evaporative coolers for vehicles have been marketed since the 1930s, but have not been commercially successful.

In the United States, cooling systems in residential and commercial buildings consume about 15% of the total heating and cooling load. On average, the air conditioning percentage is about 8%, and refrigeration is about 7%. The cooling energy use is highly dependent on location and the specific cooling application. For example, in supermarkets, refrigeration is 20–50% of the total energy use, and 1/2 of that is the energy consumed by the compressor.

Air conditioning today is concentrated in a small number of countries, but air conditioner sales are rising rapidly in emerging economies, and air conditioners are rapidly becoming the largest energy consumer in the developing world. For example, in New Delhi about half of peak electricity is consumed by air conditioners. Since only about 5% of residences in India have air conditioners, the energy consumption for residential cooling in the developing world is expected to greatly increase. China, India, and Indonesia will together account for about half of the projected increase.

A rapidly growing application of cooling is the thermal management of data centers. A data center is a building used to house the computing infrastructure for information technology operations. Data centers include the computing system software and hardware used for information management, storage, and Internet communications. Since all of the electrical energy used by computers is converted to heat, the energy consumption of a data center is very large, at least one hundred times that of an office building for the same footprint.

As a result of their large energy consumption, data centers that can use outside air or evaporative cooling technologies have lower installation and operating costs than mechanically cooled facilities. Many global service providers are building large-scale data centers in cold locations such as Finland or Sweden to cut power and cooling costs. In these locations, cold outside air is used directly as a heat sink. In cold locations with access to a river or seawater, evaporative cooling techniques have also been chosen for the thermal management of the data center computers.

The adoption of air conditioning in buildings and vehicles has had three major positive effects. People are more productive when they are cool than when they are overheated and sweaty. Thus, air conditioning has contributed to the large increase in US manufacturing productivity observed after 1930. Air conditioning has also allowed a population and a manufacturing migration to the southeast and southwest regions of the United States which have hot, and in places, very humid climates. The economic growth of the southern states was limited until summer cooling was economically feasible. Lastly, the rates of death and illness during heat waves have also decreased with the adoption of air conditioning.

1.3   Thermodynamic Background

1.3.1   Thermodynamic Properties

In this section, we review the thermodynamics of refrigeration and air conditioning systems. The thermodynamic analyses apply the first and second laws of thermodynamics to determine the thermal performance of the cooling system cycle and its components. For the purposes of this text, a thermodynamic cycle is defined as a series of cyclic processes in which a working fluid changes from one state to another state, eventually returning to its initial state or condition. For example, in a vapor compression system, the working fluid produces a net cooling effect by being compressed, condensed, expanded, and evaporated in a cyclic process.

In order to determine the performance of cooling systems, we need to compute the changes in the state of the refrigerant or working fluid flowing through the system. The changes in the state of a working fluid in a cooling system are characterized by changes in its properties such as pressure P, specific volume v, internal energy u, temperature T, entropy s, and specific heats c_p and c_v.

The property enthalpy h is defined as

$$h = u + Pv. \tag{1.1}$$

The refrigerant in the vapor phase is modeled as an ideal gas where R is the gas constant. The ideal gas law, Eq. (1.2), constrains the allowable changes in pressure, volume, and temperature.

$$Pv = RT. \tag{1.2}$$

The specific heat at constant pressure c_p relates enthalpy and temperature changes:

$$c_p = \left(\frac{\partial h}{\partial T} \right)_p, \tag{1.3}$$

and the specific heat at constant volume c_v relates internal energy and temperature changes:

$$c_v = \left(\frac{\partial u}{\partial T} \right)_v. \tag{1.4}$$

The ratio of the specific heats is denoted by γ

$$\gamma = \frac{c_p}{c_v}. \tag{1.5}$$

1.3.2 Energy Equation, Heat, and Work

An open thermodynamic system approach is used to calculate the flow of mass and energy through the components of air conditioning and refrigeration systems, such as fans, pumps, evaporators, compressors, pipes, and ducts. Analyses of these components will employ mass conservation, and the first and second laws of thermodynamics.

The mass conservation equation for a steady flow, steady state control volume is

$$\dot{m}_i = \dot{m}_e. \tag{1.6}$$

The first law energy balance equation for a steady flow, steady state control volume, shown schematically in Fig. 1.1, is

$$\dot{Q} - \dot{W} = \sum_e \dot{m}\left(h + \frac{U^2}{2} + gz\right) - \sum_i \dot{m}\left(h + \frac{U^2}{2} + gz\right). \tag{1.7}$$

In the mass and energy control volume equations, $\dot{m}$ is the mass flow rate of the working fluid entering (i) or exiting (e) the control volume, $\dot{Q}$ is the heat transfer rate from the surroundings to or from the control volume, $\dot{W}$ is the rate of work transferred to or from the control volume, U is the velocity of the working fluid entering or leaving the control volume, g is the gravitational constant, and z is the elevation of the fluid entering and leaving the control volume. The sign convention used in this text is that work done by the system inside the control volume is considered positive, for example turbine expansion; and work done on the system is considered negative, such as pump compression. Conversely, positive heat transfer is into the control volume, and negative heat transfer is out of the control volume.

Fig. 1.1 Open system control volume

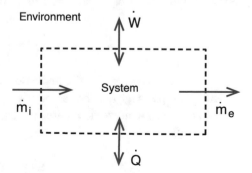

The kinetic energy and potential energy terms are usually relatively small terms in HVAC systems and thus can be neglected, in which case the first law reduces to Eq. (1.8)

$$\dot{Q} - \dot{W} = \sum_e \dot{m}h - \sum_i \dot{m}h. \tag{1.8}$$

Dividing by the flow rate $\dot{m}$, assuming a single inlet and outlet, results in an energy balance equation expressed on a per unit mass basis:

$$q - w = h_e - h_i. \tag{1.9}$$

The Gibbs differential equation relating changes in entropy to changes in enthalpy and pressure is

$$T ds = dh - v dP. \tag{1.10}$$

If a change of state is isentropic, $ds = 0$, and

$$dh = v dP. \tag{1.11}$$

Upon integration from state 1 to state 2 along an isentropic path,

$$h_2 - h_1 = \int_1^2 v dP = \bar{v}(P_2 - P_1) \tag{1.12}$$

where $\bar{v}$ is the average specific volume during the process from state 1 to state 2. For an isentropic compression of an incompressible fluid from P_1 to P_2, the energy equation, Eq. (1.8), becomes

$$- \dot{W}_s = \dot{m}\bar{v}(P_2 - P_1). \tag{1.13}$$

For gases such as refrigerant vapor, the specific volume is not constant during compression, and the ideal gas law is used to estimate the final state. The Gibbs equation for an ideal gas is

$$ds = c_p \frac{dT}{T} - R \frac{dP}{P}. \tag{1.14}$$

For an isentropic process from state 1 to 2_s, Eq. (1.14) upon integration is

$$\frac{T_{2s}}{T_1} = \left(\frac{P_2}{P_1} \right)^{\frac{\gamma-1}{\gamma}}. \tag{1.15}$$

The pressure ratio P_2/P_1 of compressors used for refrigeration and air conditioning is generally small enough that the gas may be assumed to have constant specific heat. It follows then that the isentropic work required per unit mass of gas to compress the gas from P_1 to P_2 is given by

$$w_{1-2s} = c_p(T_1 - T_{2s}) = -c_p T_1 \left[(P_2/P_1)^{(\gamma-1)/\gamma} - 1 \right]. \tag{1.16}$$

In deriving Eq. (1.16), it was tacitly assumed that the change in kinetic energy across the compressor was negligible compared to the change in enthalpy, an assumption usually valid in practice. If the pressure ratios are sufficiently large, so that the refrigerant vapor does not follow the ideal gas equation, then property tables are required.

Actual compression processes are not isentropic, and the working fluid exits with a higher temperature and entropy than the corresponding isentropic process from state 1 to state 2_s.

The adiabatic efficiency η_c of a compressor is defined as the isentropic work required to compress the gas over the specified pressure ratio divided by the actual work required to compress the gas over the same pressure ratio:

$$\eta = \frac{\dot{W}_s}{\dot{W}} = \frac{w_s}{w}. \tag{1.17}$$

1.4 Psychrometrics

1.4.1 Properties of Air–Water Vapor Mixtures

In calculating air conditioning and refrigeration system performance, we also need to compute the changes in the state of a two-component working fluid such as an air–water vapor mixture used in evaporative cooling or a lithium bromide–water mixture used in absorption cooling. Accounting for the water vapor in an air stream is very important for two reasons. First, the energy required to add or remove water vapor is non-trivial, and secondly, it has been found that there is a relatively narrow humidity range that provides acceptable thermal comfort. Energy transfers to/from an air stream with no change in the moisture content are denoted as "sensible" heat transfer, and energy transfers due to evaporation or condensation of the water vapor from a mixture are denoted as "latent" heat transfer.

There are a number of parameters that are used to quantify the properties of air–water vapor mixtures. These include the relative humidity, the humidity ratio, the dry and wet bulb temperature, the dew point temperature, the enthalpy, and the specific volume.

The relative humidity ϕ is the ratio of the mass of water vapor in an air stream relative to the maximum mass that it can contain. It is zero for dry air, and equal to one for saturated air. In terms of ideal gas pressures, the relative humidity is also expressed as the ratio of P_v, the partial pressure of the water vapor, to P_{sat}, the water vapor saturation pressure, at the same temperature. If water vapor is added to a saturated air–water vapor mixture, it will simply condense out:

$$\phi = \frac{m_v}{m_{sat}} = \frac{P_v}{P_{sat}}. \tag{1.18}$$

The humidity ratio ω is the ratio of the mass of water vapor (m_v) in an air stream to the mass of dry air (m_a):

$$\omega = \frac{m_v}{m_a}. \tag{1.19}$$

For space cooling applications, the humidity ratio is typically 0.01–0.02 kg_w/kg_a. Using the ideal gas law $PV = mRT$, we can develop equations relating ω and ϕ. If M_v is the molecular mass of water vapor, M_a is the molecular mass of air, R_u is the universal gas constant, and $P = P_v + P_a$ is the total pressure, then

$$
\begin{aligned}
\omega &= \frac{P_v\, M_v\, V/R_u T}{P_a\, M_a\, V/R_u T} \\[2mm]
&= \frac{P_v\,(18.015)}{P_a\,(28.965)} \\[2mm]
&= 0.622\frac{P_v}{P_a} \\[2mm]
&= 0.622\frac{P_v}{P - P_v} = 0.622\frac{\phi P_{sat}}{P - \phi P_{sat}}
\end{aligned}
\tag{1.20}
$$

and

$$
\phi = \frac{\omega P}{(0.622 + \omega)P_v}.
\tag{1.21}
$$

The dew point temperature T_{dp} is the temperature at which water vapor will start to condense out of a gas mixture when cooled at a constant vapor pressure, as shown on the T-s diagram in Fig. 1.2. The humidity ratio is constant during this process since the partial pressures of the air and water are constant, and the relative humidity will increase, since the mixture is being cooled.

A useful curve-fit relating vapor saturation temperature T (°C) and pressure P_v (kPa) is

$$
\ln P_v = a - \frac{b}{T + c}
\tag{1.22}
$$

Fig. 1.2 Constant pressure and adiabatic saturation processes

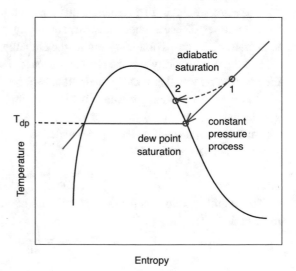

where a $= 16.6536$, b $= 4030.18$, and c $= 235$.

The total enthalpy H of an air–vapor mixture is the sum of the air (a) and vapor (v) enthalpies,

$$H = \sum m_i h_i = m_a h_a + m_v h_v. \tag{1.23}$$

The specific enthalpy of an air–vapor mixture is defined as the total enthalpy divided by the mass of dry air (kJ/kg$_a$),

$$
\begin{aligned}
h &= \frac{H}{m_a} \\
&= h_a + \frac{m_v}{m_a} h_v \\
&= h_a + \omega h_v.
\end{aligned}
\tag{1.24}
$$

For the relatively small temperature differences that occur in air conditioning applications, the specific heat of air $c_{p_a} = 1.004$ kJ/kg-K, the specific heat of water vapor $c_{p_v} = 1.868$ kJ/kg-K, and the specific heat of liquid water $c_{p_f} = 4.18$ kJ/kg-K are assumed to be constant. The enthalpy is also assumed to vary linearly with temperature, relative to a reference condition, typically 0°C. The reference condition for the enthalpy of water vapor is the liquid state at 0°C. The enthalpy of vapor thus includes the enthalpy of vaporization $h_{fg} = 2501.3$ kJ/kg. Therefore, the air and water vapor enthalpies can be expressed as

$$
\begin{aligned}
h_a &= c_{p_a} T \\
h_v &= h_{fg} + c_{p_v} T
\end{aligned}
\tag{1.25}
$$

and the specific enthalpy (kJ/kg$_a$) of an air–vapor mixture can be written as a function of temperature and humidity ratio:

$$
\begin{aligned}
h &= c_{p_a} T + \omega h_v \\
&= c_{p_a} T + \omega (h_{fg} + c_{p_v} T).
\end{aligned}
\tag{1.26}
$$

For example, the enthalpy of an air–vapor mixture at a temperature of 25°C and a humidity ratio of 0.0090 is

$$
\begin{aligned}
h &= 1.004T + \omega[2501.3 + 1.868T] \\
&= 1.004(25) + 0.0090[2501.3 + 1.868(25)] \\
&= 25.10 + 22.93 = 48.03 \text{ kJ/kg}_a.
\end{aligned}
\tag{1.27}
$$

1.4.2 Adiabatic Saturation and Wet Bulb Temperatures

Since the relative and absolute humidity of an air stream are difficult to measure directly, they are determined indirectly through temperature measurements of an unsaturated and a saturated air stream. Two processes are used, adiabatic saturation and wet bulb temperature measurements.

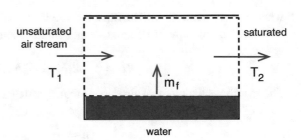

Fig. 1.3 Adiabatic saturation process

In the adiabatic saturation process, shown schematically in Fig. 1.3, unsaturated ambient air at a dry bulb temperature of T_1 and humidity ratio ω_1 is drawn into a long channel containing a pool of water. As the air flows over the water, there is evaporation of water vapor from the pool to the air stream until the air stream is saturated. The exiting air temperature T_2 is lower than the entering air temperature T_1, due to the internal heat transfer from the air to the pool of water. The adiabatic saturation temperature is the temperature resulting when the air stream becomes saturated with water vapor adiabatically, with no heat transfer to the surroundings. Make-up water is added to the pool of water at the same rate as the evaporation rate, and at temperature T_2. The total pressure P increases during this process, since water vapor is being added to the air stream, as indicated on the T-s diagram of Fig. 1.2.

We can determine the ambient humidity ratio and relative humidity using the steady state mass and energy equations applied to the control volume surrounding the wetted channel of Fig. 1.3. The mass balance equation is

$$\dot{m}_{v_1} + \dot{m}_f = \dot{m}_{v_2}$$
$$\dot{m}_a \omega_1 + \dot{m}_f = \dot{m}_a \omega_2 \tag{1.28}$$
$$\dot{m}_f = \dot{m}_a (\omega_2 - \omega_1)$$

where $\dot{m}_f$ is the rate of evaporation into the moist air stream, and $\dot{m}_a$ is the air mass flow rate through the channel.

The energy balance equation, since there is no heat or work transfer, is

$$\dot{m}_a (h_2 - h_1) = \dot{m}_f h_f = \dot{m}_a (\omega_2 - \omega_1) h_f$$
$$h_2 - h_1 = (\omega_2 - \omega_1) h_f. \tag{1.29}$$

Since $h = c_p T + \omega h_v$, upon solution for ω_1,

$$\omega_1 = \frac{c_{p_a}(T_2 - T_1) + \omega_2 (h_{v_2} - h_f)}{h_{v_1} - h_f}. \tag{1.30}$$

The air temperature measured by a thermometer with its bulb directly exposed to the air flow is defined as the dry bulb temperature T_{db}. The wet bulb temperature T_{wb} is measured by a thermometer with a wetted wick covering the bulb, so as air flows over the wetted wick, the water in the wick will evaporate. The process is very similar to the adiabatic saturation process. The energy for evaporation is provided by the air and the thermometer, so the wetted bulb will cool until it reaches a steady state temperature, defined as the thermodynamic wet bulb, where the rate of evaporation is equal to the convective heat transfer from the air stream.

Air at a dry bulb temperature of $T_{db} = 25\,^\circ$C and 100 % relative humidity ($\phi = 1$) will have a wet bulb temperature $T_{wb} = 25\,^\circ$C, since the saturated moist air mixture cannot hold any additional moisture. However, if the relative humidity is lowered below $\phi = 1$, evaporation from the wet bulb thermometer into the air–water vapor mixture can take place, and the wet bulb temperature will decrease to a value below $25\,^\circ$C.

For air–water vapor mixtures at atmospheric pressures, the wet-bulb temperature T_{wb} can be considered to be approximately equal to the adiabatic saturation temperature T_2, and used in the above mass and energy balance equations to calculate the humidity ratio and relative humidity of an air stream.

Example 1.1 Moist Air Thermodynamic Properties

The dry bulb and wet bulb temperatures of an air stream at atmospheric pressure are measured in an adiabatic saturation process to be $T_1 = T_{db} = 25\,^\circ$C and $T_2 = T_{wb} = 15\,^\circ$C, respectively. What is the absolute humidity ω_1, the relative humidity ϕ_1, and the enthalpy h_1 of the air stream?

Solution

We first determine vapor pressures and enthalpies: The vapor pressure at $T_1 = 25\,^\circ$C is

$$\ln P_{v_1} = 16.6536 - \frac{4030.18}{25 + 235} = 1.1529$$

$$P_{v_1} = 3.17\,\text{kPa}$$

and the vapor pressure at 15°C is

$$\ln P_{v_2} = 16.6536 - \frac{4030.18}{15 + 235} = 0.5328$$

$$P_{v_2} = 1.71\,\text{kPa}.$$

The make-up water enthalpy h_f is

$$h_f = c_{p_f} T_2 = (4.18)(15) = 62.7\,\text{kJ/kg-K}.$$

The vapor enthalpies are

$$h_{v_1} = h_{fg} + c_{p_v} T_1$$
$$= 2501 + (1.868)(25) = 2548\,\text{kJ/kg-K}$$

$$h_{v_2} = h_{fg} + c_{p_v} T_2$$
$$= 2501 + (1.868)(15) = 2529\,\text{kJ/kg-K}.$$

Therefore

$$\omega_2 = 0.622 \left[\frac{P_{v_2}}{P - P_{v_2}} \right]$$
$$= 0.622 \left[\frac{1.71}{101.3 - 1.71} \right] = 0.01065\,\text{kg/kg}$$

and

$$\omega_1 = \frac{c_{p_a}(T_2 - T_1) + \omega_2(h_{v_2} - h_f)}{h_{v_1} - h_f}$$
$$= \frac{(1.005)(15 - 25) + (0.01065)(2529 - 62.7)}{2548 - 62.7} = 0.00653\,\text{kg/kg}$$

$$\phi = \frac{\omega P}{(0.622 + \omega) P_v}$$
$$= \frac{(0.00653)(101.3)}{(0.622 + 0.00652)(3.17)} = 0.332$$

$$h_1 = c_{p_a} T + \omega_1 h_{v_1}$$
$$= (1.005)(25) + (0.00653)(2548) = 41.8\,\text{kJ/kg}.$$

1.4.3   Psychrometric Chart

The psychrometric chart is very useful for determining the heating and cooling energy requirements for air conditioning applications. It is a plot with (see Fig. 1.4) the dry bulb temperature on the horizontal axis, and the humidity ratio on the vertical axis. Also plotted on the chart are other moist air properties, namely lines of constant relative humidity, constant wet bulb temperature, constant enthalpy, and constant specific volume. The vertical axis is placed on the right-hand side for ease of use. Note that the saturation line $\phi = 1$ defines the upper extent of the chart.

All of the moist air thermodynamic properties are drawn on the psychrometric chart using the equations derived in the previous section, so given one set of conditions, such as inlet air from the environment, one can visualize and quickly calculate the changes in the thermodynamic state of the air stream that are required to meet a set of conditions for an occupied space, such as thermal comfort and ventilation.

As shown in Fig. 1.5, the lines of constant enthalpy slope downward, since as the dry bulb temperature increases, the humidity ratio decreases to maintain a constant enthalpy. The diagonal lines of constant wet bulb temperature also slope downward, coinciding with the dry bulb temperature at the saturation line. The lines of constant enthalpy and wet bulb temperature are almost parallel to each other, since the measurement of wet bulb temperature approximates a constant enthalpy process.

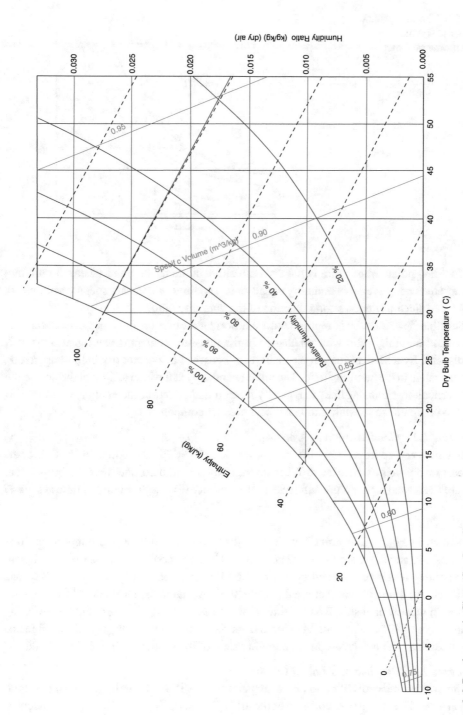

Fig. 1.4 Psychrometric Chart (Sea Level)

Fig. 1.5 Thermodynamic
Properties on the
Psychrometric Chart

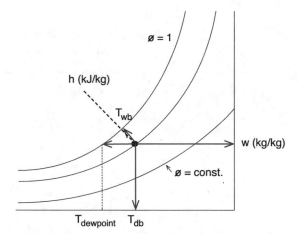

The dew point temperature is the dry bulb temperature at the point where the mixture is saturated and the relative humidity equals one. The lines of constant specific volume are very steep since v is not a strong function of the humidity ratio.

Sensible heat transfer is represented by a horizontal line on the chart, as it occurs at a constant humidity ratio with no change in the moisture content. Latent heat transfer is represented by a vertical line on the chart, as it occurs at a constant dry bulb temperature, with a change in the humidity ratio. The mixture enthalpy at the intersection of the horizontal and vertical heat transfer lines is typically labeled as h_x, and in the energy equation, it is used to compute the sensible and latent heat transfer components.

Example 1.2   Condensation of Moist Air

Moist air is cooled from $T_1 = 40°C$ and $\phi_1 = 0.30$ to $T_2 = 30°C$ and then to 15 °C. Plot the process on a psychrometric chart. What are the humidity ratio and the wet bulb temperature at state 1? At what temperature will condensation occur? What is the humidity ratio at state 4?

Solution

As shown in Fig. 1.6, at point 1 with $T_1 = 40°C$ and $\phi_1 = 0.30$, the humidity ratio is $\omega_1 = 0.0140$ and the wet bulb temperature is 25.1 °C. The cooling process from 1-2 is sensible cooling and is represented by a horizontal line at a constant humidity ratio. At state 2, the relative humidity has increased to $\phi_2 = 0.53$. As the moist air is cooled further, condensation will occur at state 3 on the saturation line $\phi_3 = 1$ with a corresponding dew point temperature of 19 °C. The cooling process from state 3 to state 4 includes both sensible and latent cooling with a decrease in the humidity ratio of the saturated air to $\omega_4 = 0.0106$.

Example 1.3   Sensible and Latent Cooling

An air conditioner cools 35 °C dry bulb temperature and 35% relative humidity air to saturation at 7 °C. The volumetric flow rate of the air is $\dot{V} = 120$ L/s (dry). How much condensed water must be drained per hour from the air conditioner? What is the sensible and the latent

Fig. 1.6 Condensation of an air–water vapor mixture. (Example 1.4.3)

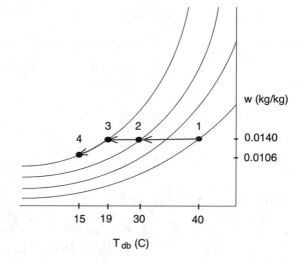

heat transfer from the air stream?

Solution

As shown in Fig. 1.7, at point 1, the humidity ratio $\omega_1 = 0.012$, the enthalpy $h_1 = 66.5$ kJ/kg, and specific volume $v_1 = 0.89$ m^3/kg. At point 2, the humidity ratio $\omega_2 = 0.0063$, the enthalpy $h_1 = 22.5$ kJ/kg, and at point x, the enthalpy $h_x = 51.0$ kJ/kg. The mass flow rate of the air stream is

$$\dot{m}_a = \frac{\dot{V}}{v}$$
$$= 0.120/0.89 = 0.135 \, \text{kg/s}.$$

Fig. 1.7 Sensible and Latent Cooling. (Example 1.4.3)

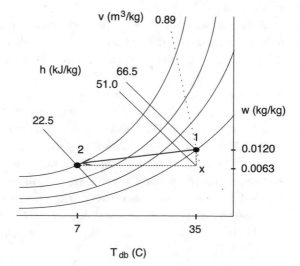

From the continuity equation for the condensed water m_w,

$$\dot{m}_w = \dot{m}_{v1} - \dot{m}_{v2} = \dot{m}_a(\omega_1 - \omega_2)$$
$$= (0.135)(0.012 - 0.0063)(3600 \text{ sec/hr})$$
$$= 2.77 \text{ kg/hr.}$$

The sensible heat transfer $\dot{Q}_s$ is

$$\dot{Q}_s = \dot{m}_a(h_x - h_2)$$
$$= (0.135)(51.0 - 22.5) = 3.85 \text{ kW}$$

and the latent heat transfer $\dot{Q}_l$ is

$$\dot{Q}_l = \dot{m}_a(h_1 - h_x)$$
$$= (0.135)(66.5 - 51.0) = 2.09 \text{ kW.}$$

1.5   Thermal Comfort

The primary function of an HVAC cooling system is to provide acceptable environmental conditions for the occupants. Accordingly, the air temperature of the indoor space needs to be chosen by the building engineer or user so that most of the occupants feel comfortable most of the time independent of the outdoor environmental conditions.

Based on extensive thermal comfort research, ASHRAE (2017b) has developed a set of recommendations for thermal comfort that include parameters such as the dry bulb temperature, mean radiant temperature, relative humidity, and indoor air speed. The ranges and combinations of these parameters that provide comfortable thermal conditions for people have been measured in controlled conditions for many different clothing levels and ranges of metabolic activity.

A thermal sensation scale ranging from cold to hot is used in thermal comfort research to quantify a person's thermal sensation. The scale has integer values: $+3$ for feeling hot, $+2$ for warm, $+1$ for slightly warm, 0 for neutral, -1 for slightly cold, -2 for cool, and -3 for feeling cold. The average of the responses from the subjects in the experiments is termed the Predicted Mean Vote (PMV), and the Predicted Percentage Dissatisfied (PPD) is defined as the percentage of people in a group that would find the thermal environment unacceptable. The ASHRAE recommendations are based on a 90% acceptance rate, based on a PMV between slightly cool and slightly warm.

For a neutral thermal sensation, from the First Law of thermodynamics, the rate of energy generation by a person's metabolism must equal the rate of energy transferred to the environment and the rate of work. The thermal energy is transferred by sensible and latent heat transfer from the exposed skin and the clothing layers, respiration, and work. The sensible heat transfer is a combination of convection and radiation from the skin and the

Table 1.1 Metabolic rates for typical tasks

Activity	M (mets)
Sleeping	0.7
Seated, quiet	1.0
Light office work	1.1
Standing, relaxed	1.2
Walking about	1.7–2.1
Light machine work	2.0
Lifting/packing	2.1
Calisthenics/exercise	3.0–4.0

Source Adapted from *2017 ASHRAE Handbook—Fundamentals*, Chap. 9

surface of the clothing. For a warm or hot sensation, the metabolic energy and transferred energy do not balance, and as a result, energy is stored in the body and there is an increase in the skin and body temperature. Conversely, a cool or cold thermal sensation results from a decrease in the skin and body temperature.

The physical activity level of a person is quantified by the metabolic rate parameter M with units of $mets$. A value of 1.0 met is defined as the metabolic rate of a sedentary person and is equal to 58.2 W/m^2. For a surface area of 1.8 m^2, this is equal to 105 W. At rest, about 90% of the energy transfer is from the skin and clothing, and 10% respiration. As the activity level increases, the respiration fraction increases significantly. The metabolic rates for various activity levels are given in Table 1.1.

The clothing insulation level of a person is quantified by the parameter I with units of clo, a measure of thermal resistance. A value of 1.0 clo is equivalent to 0.155 (m^2-K)/W. Various clothing insulation values for a person that is not moving are given in Table 1.2. For example, a typical summer clothing level of trousers and a short sleeve shirt has a clo value of 0.57. The clothing insulation values for an active person who is moving decrease due to the air motion between and through the clothing layers. An estimate of the clothing insulation for an active person with $1.2 < M < 2.0$ is

$$I_{active} = I \times (0.6 + 0.4/M). \tag{1.31}$$

Equation (1.32) is used to calculate the optimal air temperature in a space for people with a given clothing level I, an activity level above 1.2 mets, for an air speed less than 0.15 m/s and 50% relative humidity. As the clothing and activity levels increase, the set point of the thermostat will need to be lowered to maintain acceptable thermal comfort. In general, the air temperature needs to be decreased by 0.6 °C for every 0.1 clo increase in clothing, and also decreased by 3 °C per met increase in activity above 1.2 met:

$$T_{a,opt} = 27.2 - 5.9I - 3.0(1 + I)(M - 1.2). \tag{1.32}$$

Table 1.2 Clothing thermal resistance values

Garment description	I (clo)
no clothes	0
shorts and T-shirt	0.36
knee-length skirt, short-sleeved shirt, panty hose, sandals	0.54
trousers and short-sleeved shirt	0.57
trousers and long-sleeved shirt	0.61
sweat pants and sweat shirt	0.74
same as above, plus suit jacket	0.96

Source Adapted from *2017 ASHRAE Handbook—Fundamentals*, Chap. 9

The ASHRAE thermal comfort zones are defined in terms of an operative temperature T_o which is a measure of the combined effect of convection and radiation. For situations where the heat transfer coefficients for radiation and convection are approximately equal, and the local air speed is less than 0.2 m/s, the operative temperature is given by

$$T_o = \frac{T_a + T_r}{2} \tag{1.33}$$

where T_a is the average temperature of the air in the space, and T_r is the mean radiant temperature, i.e., the average temperature of the surrounding surfaces, including windows, walls, ceiling, and floor. In the interior zones of commercial buildings, the operative temperature is essentially equal to the air temperature. In perimeter zones, the effect of windows will increase the influence of thermal radiation on the operative temperature.

The indoor local air speed V is also an important consideration to avoid local discomfort due to draft. For seated office activities, with operative temperatures above 25.5 °C, the air speed should not exceed 0.8 m/s, and for operative temperatures below 22.5 °C, the air speed should not exceed 0.15 m/s. For operative temperatures between these limits, a curve-fit for the maximum air speed is

$$V = 50.49 - 4.4047\,T_a + 0.096425\,T_a^2 \tag{1.34}$$

where V is in m/s, and T_a is in °C.

Example 1.4 Thermal Comfort

A commercial space is being remodeled from an office to a physical therapy facility. What is the change in $T_{a,opt}$, the optimum operative air temperature?

Solution

Assume that the office staff wear pants and long sleeve shirts with a stationary insulation value of 0.6 clo, and is a mixture of seated and standing people with an average activity of

1.2 mets. For the physical therapy facility, assume that the staff and clients wear T-shirts and shorts with an insulation value of 0.22 clo (moving), and an average activity level of 3.0 mets. The optimum air temperature for the office is

$$T_{a,opt} = 27.2 - 5.9I - 3.0(1+I)(M-1.2)$$
$$= 27.2 - 5.9(0.6) = 23.7\,\text{C}$$

and the optimum air temperature for the physical therapy facility is

$$T_{a,opt} = 27.2 - 5.9I - 3.0(1+I)(M-1.2)$$
$$= 27.2 - 5.9(0.22) - 3.0(1+0.22)(3.0-1.2) = 19.3\,\text{C}.$$

The air conditioning control system will need to be reprogrammed to lower the thermostat set point in the space from 23.7 °C to 19.3 °C.

The ASHRAE thermal comfort zones are presented in Fig. 1.8, which is an overlay of the comfort zones on the psychrometric chart with the operative temperature on the horizontal axis and the humidity ratio on the vertical axis. Two zones are shown, one for 0.5 clo of clothing insulation, and one for 1.0 clo of insulation. As discussed above, these insulation levels are the lower and upper ranges of clothing insulation typically worn in office environments. The comfort zone will move to regions of lower temperature with an increase in clothing insulation, metabolic rate, and mean radiant temperature. Conversely, the comfort zone will move to regions of higher temperature with a decrease in clothing insulation, metabolic rate, and mean radiant temperature.

The upper humidity limit is 0.012 kg$_v$/kg$_a$, as above that level, one's skin feels damp and uncomfortable. There are no established lower humidity limits for thermal comfort, however, non-thermal comfort factors such as skin and eye dryness should be considered.

Fig. 1.8 Acceptable ranges of temperature and humidity for summer thermal comfort. (Adapted from ASHRAE 2017b)

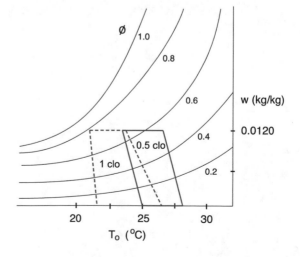

In Fig. 1.8, the local air speed in the zone is assumed to be not greater than 0.2 m/s. Elevated air speed can be used to increase the maximum operative temperature using Eq. (1.34).

The approximate middle of the range, 25 °C and 50% relative humidity, is widely used as a cooling design condition for the 0.5 clo comfort zone.

1.6   Further Reading

Additional background information about cooling technologies is provided in McQuiston (2004), Pita (2002), Kuehn et al. (1998), and Dossat and Horan (2002). A recommended handbook about industrial refrigeration systems is Stoecker (1998).

References

ASHRAE, (2017a) ASHRAE Handbook – Fundamentals, Chapter 14 –Climatic Design. ASHRAE, Atlanta, GA

ASHRAE, (2017b) ASHRAE Handbook – Fundamentals, Chapter 9 –Thermal Comfort. ASHRAE, Atlanta, GA

Dossat R, Horan T (2002) Principles of Refrigeration. Prentice Hall, Hoboken, New Jersey

Gordon R (2016) The Rise and Fall of American Growth. Princeton University Press, Princeton, New Jersey

Kuehn T, Ramsey J, Threlkeld J (1998) Thermal Environmental Engineering. Prentice Hall, Hoboken, New Jersey

McQuiston F, Parker J, Spitler J (2004) Heating, Ventilating, and Air Conditioning. John Wiley, New York

Pita E (2002) Air Conditioning Principles and Systems. Prentice Hall, Hoboken, New Jersey

Stoecker W (1998) Industrial Refrigeration Handbook. McGraw Hill, New York

Vapor Compression Cooling Cycles

2

2.1 Introduction

In this chapter, we examine vapor compression cycles for refrigeration and air conditioning. We first review the Carnot cycle, then perform a thermodynamic analysis of basic vapor compression cycles to illustrate the various thermal efficiency metrics and the effect of condenser and evaporator temperatures. We then discuss the thermodynamic properties of common refrigerants, compute part load performance, and finish with an analysis of multistage vapor compression cooling cycles. With increasing climate change due to the buildup of greenhouse gas emissions in the atmosphere, there has been a worldwide impetus to transition to refrigerants that have a low or even zero global warming potential.

Since the second law of thermodynamics states that thermal energy can flow only from high temperatures to low temperatures, a thermodynamic cycle is required to produce a cooling or refrigeration effect that results in a net flow of thermal energy originating from a low-temperature object or space to be cooled and then transferred to a higher temperature heat sink. A vapor compression cycle takes advantage of the phase change of a working fluid in an evaporator and a condenser to effect the net heat transfer from a low-temperature source to a high-temperature sink.

2.2 Carnot Refrigeration Cycle

The scientific theory of thermodynamic cycles was first developed by Sadi Carnot (1796–1832), a French engineer, in 1824. His theory has two main axioms. The first axiom is that in order to use a flow of energy to generate power, there need to be two bodies at different temperatures, a hot body and a cold body. Work is extracted from the flow of thermal energy from the hot to the cold body or reservoir. The second axiom is that there must be at no

© The Author(s), under exclusive license to Springer Nature Switzerland AG 2023
A. T. Kirkpatrick, *Introduction to Refrigeration and Air Conditioning Systems*,
Synthesis Lectures on Mechanical Engineering,
https://doi.org/10.1007/978-3-031-16776-8_2

point a useless flow of energy, so heat transfer at a constant temperature is needed. Carnot developed an ideal heat engine cycle, which is reversible, i.e., if the balance of pressures is altered, the cycle of operation is reversed.

The efficiency of this cycle, known as the Carnot cycle, is a function only of the reservoir temperatures. The Carnot cycle, since it is reversible, is the most efficient possible, and thus is the standard to which all real cycles are compared. An ideal refrigeration cycle is a Carnot cycle operated in reverse. It consumes work during the compression process, and has the net effect of removing thermal energy from the low-temperature reservoir at temperature T_l, and rejecting it to the high-temperature reservoir at temperature T_h.

The state processes for the ideal refrigeration cycle are plotted in Fig. 2.1. The four basic processes are

> 1 to 2   isentropic compression
> 2 to 3   constant temperature heat rejection, Q_h
> 3 to 4   isentropic expansion
> 4 to 1   constant temperature heat absorption, Q_l

One efficiency metric for cooling cycles is the Coefficient of Performance (COP), which is the thermal energy transferred from the object to be cooled, Q_l, divided by the compressor work W required. The COP is the dimensionless ratio of how much heat is transferred out of the cooled space relative to the amount of work that is used to accomplish this task. Note that the COP is larger than unity:

$$COP = \frac{Q_l}{W} = \frac{Q_l}{Q_h - Q_l} = \frac{T_l}{T_h - T_l}. \tag{2.1}$$

The Carnot cycle COP increases as the temperature difference between the high and low temperatures decrease, and it increases as the temperature of the low-temperature body T_l increases.

Fig. 2.1 The Carnot refrigeration cycle

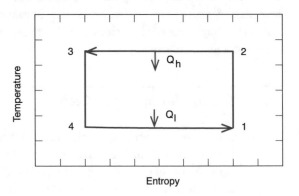

2.3   Vapor Compression Cycle Operation

The basic components of the vapor compression cycle are shown in Fig. 2.2. A compressor or pump is used to compress the vapor from low to high pressure, heat energy Q_h is transferred to the environment in the condenser, an expansion valve is used to reduce the pressure, and thermal energy Q_l is transferred from the space or object to be cooled to the evaporator.

The ideal vapor compression cycle operates in a similar fashion to the ideal Carnot refrigeration cycle, but with constant pressure condensation and evaporation of the refrigerant or working fluid. Representative state points on the temperature–entropy (T-s) diagram are plotted in Fig. 2.3. The four basic processes in the cycle in Fig. 2.3 are

1 to 2   vapor compression
2 to 3   constant pressure condensation
3 to 4   adiabatic expansion
4 to 1   constant pressure evaporation

Most of the heat transfer at the lower and higher temperatures is accomplished isothermally, similar to the Carnot cycle. As we will see, the overall performance of a vapor compression system is very dependent on the saturation temperatures of the refrigerant in the evaporator and the condenser.

The vapor compression cycle can also be plotted on a pressure–enthalpy (P-h) plot, as shown in Fig. 2.4, so that changes in enthalpy from one state to another can be read directly from the diagram.

2.4   Vapor Compression Cycle Analysis

We use energy balances on each of the components shown in Fig. 2.2 to determine the overall system performance. For each component, the open system energy equation per unit mass of refrigerant is

Fig. 2.2 Vapor compression cycle components

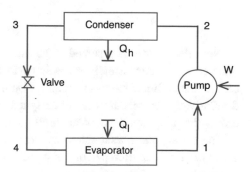

Fig. 2.3 T-s diagram for ideal
vapor compression cycle

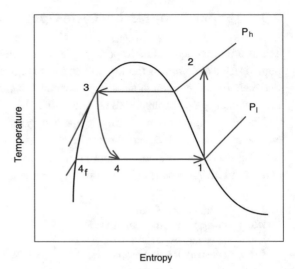

Fig. 2.4 Pressure–Enthalpy
states of vapor compression
cycle

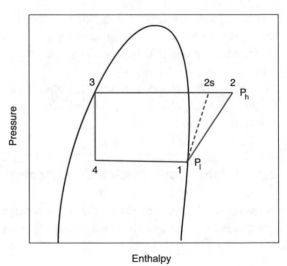

$$q - w = \Delta h. \tag{2.2}$$

Note that state 1 is assumed to be saturated vapor, and state 3 is saturated liquid. In practice, evaporators usually are designed to superheat the refrigerant by a small amount to ensure that no liquid enters the compressor, thereby avoiding possible compressor damage. Similarly, condensers are usually designed to subcool the refrigerant a certain amount to ensure that only liquid enters the throttling device for optimum expansion performance.

Starting at point 1, a saturated vapor at low pressure P_l is compressed to a higher pressure P_h and a superheated temperature at state 2. The energy equations for isentropic (s) and actual compression (c) processes are

Fig. 2.5 Temperature–entropy
states of actual vapor
compression cycle

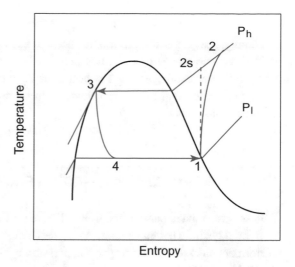

For an ideal case with adiabatic and reversible compression, the entropy of the exit
state 2s, s_{2s}, is equal to the entropy of the inlet state 1, s_1. In actual operation, there are
irreversibilities and there is an increase in entropy, and the working fluid is compressed to
a higher temperature T_2, as shown in Fig. 2.5.

The isentropic efficiency η_c of the compressor is defined as the ratio of the isentropic to
the actual work:

$$\eta_c = \frac{w_s}{w_c}. \tag{2.4}$$

The enthalpy at state 2 therefore depends on the compressor efficiency:

$$h_2 = h_1 + \frac{h_{2s} - h_1}{\eta_c}. \tag{2.5}$$

From state 2 to state 3, heat is transferred from the working fluid to the environment at
temperature T_h through constant pressure heat rejection. The heat transfer process is primarily condensation from a superheated vapor to a saturated liquid. In operation, there is a
pressure drop in the working fluid as it flows through the condenser. The heat transfer in the
condenser is given by

$$q_h = h_3 - h_2. \tag{2.6}$$

A throttling device is used to reduce the working fluid pressure in an adiabatic expansion
from the high-pressure state 3 to the low-pressure state 4. Since there is no work done by the
throttle, and the throttling is assumed to be adiabatic, the resulting energy balance equation
is

$$\begin{aligned} w_s &= h_{2s} - h_1 \\ w_c &= h_2 - h_1. \end{aligned} \tag{2.3}$$

$$h_4 = h_3. \tag{2.7}$$

Finally, heat is transferred from the body to be cooled at temperature T_b to the working fluid at temperature T_l in a constant pressure heat addition in the evaporator. This is the refrigeration or cooling effect. The heat transfer in this case is through evaporation in the two-phase region under the vapor dome from state 4 to state 1, and the resulting energy balance equation is

$$q_l = h_1 - h_4. \tag{2.8}$$

2.5   Efficiency Measures—COP, EER, and SEER

We use three related parameters, COP, EER, and SEER, to quantify the thermal efficiency of thermodynamic cooling cycles. As in the Carnot refrigeration cycle, the Coefficient of Performance (COP) for the vapor compression cycle is the refrigeration heat transfer q_l divided by the compressor work w_c required:

$$COP = \frac{q_l}{w_c}. \tag{2.9}$$

The cooling capacity is defined as the rate of cooling $\dot{Q}_l$ (kW),

$$\dot{Q}_l = \dot{m} q_l \tag{2.10}$$

where $\dot{m}$ is the working fluid mass flow rate, typically in units of kg/s. In the English system, the cooling capacity is defined in terms of tons, where 1 ton $= 12{,}000$ Btu/hr $= 3.52$ kW. The ton unit is historically based on ice storage, and is the amount of energy required to melt 2000 lbs of ice at 32 °F in 24 h.

The COP is also expressed as a ratio of the cooling capacity to the compressor power:

$$COP = \frac{\dot{m} \, q_l}{\dot{m} \, w_c} = \frac{\dot{Q}_l}{\dot{W}_c}. \tag{2.11}$$

The reciprocal of the COP is the kW/ton, where the cooling capacity is in tons, and the compressor power is in kW. Typical values for the COP range from 4.0 to 7.0, corresponding to kW/ton values from 0.88 to 0.50, depending on the type and size of the compressor, and the compressor and evaporator temperatures. Representative values for three types of compressors are given in Table 2.1. The chiller ratings are reported for 100% load and at 6.7 °C (44 °F) leaving chilled water temperature and 29.4 °C (85 °F) entering condenser water temperature.

The efficiencies of air conditioners are compared using the metrics EER, the Energy Efficiency Ratio, and SEER, the Seasonal Energy Efficiency Ratio. The EER is defined as the evaporator cooling capacity in BTU divided by the compressor energy input in Watt-hour, i.e., a COP with mixed units. Applying the unit conversion to the COP, we have

Table 2.1 Typical COP and kW/ton values for compressors

Compressor	COP	kW/ton
Reciprocating	4.1–4.5	0.78–0.85
Screw	4.7–5.7	0.62–0.75
Centrifugal	5.0–7.0	0.50–0.70

$$EER = 3.412 \, COP. \tag{2.12}$$

The EER is calculated assuming a constant outside temperature T_h of 35 °C, and a constant inside temperature of 27 °C. It is used primarily for small window-mounted air conditioners. Air conditioners with EER > 11.6 are given an Energy Star Rating by the US EPA. Air conditioners are currently available that have EERs as high as 16.

By way of comparison, the maximum EER is found by assuming an ideal Carnot cycle operating between an outside temperature of $T_h = 35 \, °C$ (308 K) and an inside temperature of $T_l = 27 \, °C$ (300 K):

$$EER_{max} = 3.412 \, \frac{T_l}{T_h - T_l} = 3.412 \, \frac{300}{308 - 300} = 128. \tag{2.13}$$

This is about ten times the current EER, indicating that there is plenty of room for efficiency improvements, even accounting for the temperature differences across the evaporator and condenser.

The COP is averaged over a cooling season to form the SEER, the Seasonal Energy Efficiency Ratio. Like the EER, the SEER also has mixed units of Btu/W-hr:

$$SEER = 3.412 \, \frac{1}{t} \int_0^t COP \, dt. \tag{2.14}$$

The SEER is calculated using an average value of the outdoor temperature of $T_h = 28.3 \, °C$, a range of temperature intervals from 18 to 40 °C, and a specified amount of time in each temperature interval. Since the SEER outside temperature is less than the EER outside temperature, in accordance with the Second Law, the SEER will be greater than the EER for the same equipment. As a rule of thumb,

$$SEER = 3.9 \, COP. \tag{2.15}$$

An air conditioner with a COP of 4.0 will have an EER of 13.6, and a SEER approximately equal to 15.6.

To increase energy efficiency and reduce greenhouse gas emissions from electric power plants, many governments have adopted minimum performance standards for air condition-ers. For example, as of 2015, in accordance with US EPA regulations, all split system central air conditioners in the Southeast and Southwest United States must have a minimum SEER value of 14. Typical values for DOE Energy Star central air conditioners range from 18 to 26. Measures such as the incorporation of variable speed compressors and fans are used to increase air conditioning system efficiency.

2.6   Refrigerants

There are many choices for the working fluid in vapor compression cycles. The major requirement is that the thermal properties of the refrigerant fluid match the requirements of the given application. The thermal-fluid requirements are a moderate condensation pressure so high-pressure rated components are not needed, an evaporation pressure above atmospheric pressure to prevent air leakage into the system, a boiling temperature of about 5–10 °C at atmospheric pressure, a high critical temperature so phase change heat transfer can be employed, a high latent heat of vaporization to reduce the required working fluid flow rate, high thermal conductivity, low kinematic viscosity, and a low freezing temperature.

Other important considerations are low global warming potential, toxicity, stability, flammability, cost, lubricant and materials compatibility, and environmental impact. The refrigerant properties are also a consideration in compressor selection, as reciprocating compressors are suitable for working fluids with a low specific volume, and centrifugal compressors are best adapted to working fluids with high specific volumes. Refrigerants also differ with respect to their miscibility with oil, and also the amount of water that they can hold in solution. As this section discusses in detail, there is no "perfect" refrigerant, and the choice for a particular application will necessarily involve compromises.

The majority of refrigerants in use at the present time in cooling systems are halocarbons. A halocarbon is a carbon–hydrogen molecule bonded to the molecules chlorine and/or fluorine. There are four major types of halocarbons, chlorofluorocarbon (CFC), hydrochlorofluorocarbons (HCFC), hydrofluorocarbons (HFC), and hydrofluoro-olefins (HFO):

- **CFC**—fully halogenated hydrocarbon containing chlorine, fluorine, and carbon atoms, and no hydrogen atoms.
- **HCFC**—partially halogenated hydrocarbon containing chlorine, fluorine, carbon, and hydrogen atoms.
- **HFC**—partially halogenated hydrocarbon containing fluorine, carbon, and hydrogen atoms, and no chlorine atoms.
- **HFO**—partially halogenated hydrocarbon containing fluorine, carbon, and hydrogen atoms, and no chlorine atoms.

Halocarbons were first synthesized in 1928 by T. Midgley (1889–1944), a mechanical engineer at the General Motors Research Laboratory, initially for automotive air conditioning applications. They are synthesized by substituting one or more of the hydrogen atoms in a methane or ethane molecule with atoms from the halogen family of chlorine, fluorine, and/or bromine. CFC's and HCFC's are very stable, non-toxic, and have a relatively low cost.

The halocarbon numbering system developed by Midgley is as follows. The leading **R** signifies a refrigerant for vapor compression systems. For carbon-based refrigerants, the first number indicates the number of carbon atoms minus one, the second number is the number of hydrogen atoms plus one, and the third number is the number of fluorine atoms. The letter **a** specifies a specific configuration of the molecule. The refrigerant R-134a has two carbon atoms, two hydrogen atoms, and four fluorine atoms resulting in the chemical formula $C_2H_2F_4$. The refrigerant R-32, based on methane CH_4, has two of the four hydrogen atoms replaced by fluorine, resulting in the chemical formula CH_2F_2.

Inorganic compounds are numbered by adding 700 to their molecular mass. Ammonia ($M = 17.03$) is numbered R-717, carbon dioxide ($M = 44.01$) is numbered R-744, and water ($M = 18.015$) is numbered R-718.

A mixture of refrigerants is defined as a zeotropic mixture, and will have a different saturation temperature for each component at the same pressure. The temperature difference between the saturated vapor temperature and the saturated liquid temperature at constant pressure is defined as the temperature glide. The temperature glide of a single-component refrigerant is zero. For a zeotropic mixture at a given temperature or pressure, the compositions of the liquid and vapor phases are different, with the vapor composition having a higher concentration of the low boiling point components in the mixture. As a result of this composition difference, zeotropic mixtures have a temperature glide when they boil or condense.

The safety of refrigerants depends on the hazards associated with their use, such as toxicity and flammability. ASHRAE Standard 34 classifies refrigerants based both on their allowable exposure limit and their flammability. There are six toxicity and flammability classifications (A1, A2, A3, B1, B2, and B3). Group A1 refrigerants are the least hazardous, and group B3 are the most hazardous. The letter A designates refrigerants with an occupational exposure limit (OEL) ≥ 400 parts per million (ppm), and the letter B designates refrigerants with an OEL < 400 ppm. The number designation indicates flammability:

- Class 1: No flame propagation in air at $60\,°C$ and 1 atm.
- Class 2: It exhibits flame propagation at $60\,°C$ and 1 atm, lower flammability limit (LFL) >0.10 kg/m^3 at $23\,°C$ and 1 atm, and heat of combustion $<19,000$ kJ/kg.
- Class 2L: Class 2 refrigerants with a maximum burning velocity of 100 mm/s at $23\,°C$ and 1 atm.
- Class 3: It exhibits flame propagation at $60\,°C$ and 1 atm, and LFL ≤ 0.10 kg/m^3 at $23\,°C$ and 1 atm, and heat of combustion $\geq 19,000$ kJ/kg.

Chillers using flammable refrigerants in industrial and commercial applications employ gas leak detectors and continuous pressure control to ensure safe operation. Several Underwriter

Laboratories (UL) and International Electrotechnical Commission (IEC) standards limit the charge size of flammable refrigerants to 150 g in stand-alone systems.

From the 1930s to the 1990s, most refrigerants used in vapor compression cooling systems were of the CFC type, primarily Freon-12. Because of its non-toxicity, Freon became the preferred coolant in large air conditioning systems. In fact, public health codes in many American cities were revised to designate Freon as the only coolant that could be used in public buildings.

However, as a result of its stability and long lifetime in the environment, Freon is able to migrate upward to the stratospheric ozone layer and remain in the stratosphere with a half-life of about 100 years. The stratospheric ozone layer absorbs harmful ultraviolet (UV) solar radiation, reducing the amount of UV radiation that penetrates the ground level. However, the chlorine atoms in the CFC's and HCFC's react with stratospheric ozone, reducing its concentration in the stratosphere, thus resulting in an increase in the ultraviolet radiation levels on the earth's surface.

As a result of the discovery of the effect of the chlorine in CFC's on the ozone layer, the 1989 Montreal Protocol mandated the phase-out of CFC's by 1996, and a phase-out of HCFC's (-21, -22, -123) from 2004 to 2030. Since they have no chlorine atoms, HFCs have an ozone depletion potential of zero. Freon-12 was replaced by HFC-134a, first synthesized in 1936, by A. Henne (1901–1967) an American colleague of T. Midgley. Another widely used refrigerant, HCFC-22, was replaced by HFC-410a.

However, all halocarbons, including HFCs, are also greenhouse gases, and thus their manufacture and use have became increasingly regulated. A greenhouse gas is a gas which absorbs infrared radiation from the earth's surface, reducing the radiation loss from earth to space, thus increasing the temperature of the atmosphere, earth, and oceans. The metric used for the climatic impact of greenhouse gases is the global warming potential (GWP), normalized to a value of 1 for carbon dioxide. With increasing climate change due to the buildup of greenhouse gas emissions in the atmosphere, there has been a worldwide impetus to transition to refrigerants that have a low or even zero global warming potential.

The 2016 Kigali Amendment to the Montreal Protocol commits nations to a staged phaseout of HFC's. The United States and Europe will reduce HFC use to 15% of 2012 levels by 2036, and much of the rest of the world will reduce HFC use to 20% of 2021 levels by 2045, with the exception of some Middle Eastern and Asian nations in hot climates, such as Saudi Arabia, Kuwait, India, and Pakistan.

The refrigerant HFC-134a has a GWP of 1430, and HFC-410a has a GWP of 1920. HFC-410a is a mixture of 50% HFC-32 and 50% HFC-125. Since HFC-410a has a relatively high GWP, it will be phased out for use in new cooling systems in 2023. Two replacement refrigerants are the single component HFC-32, which has a GWP of 675, and the zeotropic refrigerant R-454b, which has a GWP of about 466.

A recently developed class of refrigerant are hydrofluoro-olefins (HFO), which are unsaturated hydrofluorocarbons. Some HFO refrigerants that have a very low (<1) global warming potential are HFO-1234yf, and HFO-1234ze. These HFOs are being blended with existing

Table 2.2 Comparison of refrigerant Global Warming Potential (GWP) and Safety

Refrigerant	Name	GWP	Safety
R-22	Chlorodifluoromethane	1760	A1
R-134a	Tetrafluoroethane	1430	A1
R-410a	Puron®	1920	A1
R-32	Difluoromethane	675	A2L
R-454b	Opteon® XL41	466	A2L
R-466a	Solstice® N41	733	A1
R-290	Propane	3	A3
R-600a	Isobutane	20	A3
R-744	CO_2	1	A1
R-717	Ammonia	0	B2L

Source Adapted from *2017 ASHRAE Handbook—Fundamentals*, Chapter 29

HFC's to reduce their global warming potential. The refrigerant R-513a has been formulated as a replacement for HFC-134a. It is a blend of 56% HFO-1234yf and 44% HFC-134a, and has a GWP of about 630. The refrigerant R-454b is a blend of 31% HFO-1234yf and 69% HFC-32. However, as indicated by Table 2.2, many of these newer HFO refrigerants are listed as safety Class A2L and are mildly flammable.

Naturally occurring refrigerants that have a low or zero global warming potential are propane, isobutane, carbon dioxide, and ammonia, also listed in Table 2.2. Carbon dioxide is inexpensive, a byproduct from industrial production, non-toxic, and nonflammable. Since its critical temperature is $31\,°C$, carbon dioxide has a loss of capacity at high temperatures. Due to its high condenser pressure of about 7000 kPa at $30\,°C$, a high-pressure steel pipe system is required. Note that propane and isobutane, listed as safety Class A3, are highly flammable.

Ammonia is relatively inexpensive, has a high latent heat, is environmentally stable, non-corrosive, and mildly flammable. It has the largest refrigeration effect per kg ($\simeq 1100$ kJ/kg) of refrigerant fluids. It is used in large-scale industrial food refrigeration, such as cold storage, frozen food, dairy, and meat processing. However, ammonia is highly toxic, so it is not suitable for domestic refrigeration and air conditioning. Ammonia systems require the use of stainless steel piping, since ammonia becomes corrosive to non-ferrous metals such as copper and brass when it absorbs moisture.

The thermodynamic properties of some widely used refrigerants are given in Table 2.3. Refrigerants with similar vapor pressures will evaporate and condense at the same pressures. Therefore, a vapor compression cycle designed to operate at given condenser and evaporator pressures will perform comparably for two refrigerants with similar vapor pressures. However, differences in specific volumes and latent heat will require different working fluid flow rates for each refrigerant. The bubble point, the temperature at which the first bubble of

Table 2.3 Thermodynamic properties of selected refrigerants

Label	Name	Formula	Molecular Mass (g)	$T_{Boiling}$ at 1 bar (°C)	Critical Temp. (°C)	P_{sat} at −15°C (kPa)	P_{sat} at 30°C (kPa)
R-32	Difluoromethane	CH_2F_2	52.0	−51.7	78.3	489	1927
R-134a	Tetrafluoroethane	$C_2H_2F_4$	101.06	−26.2	102	164	770
R-1234yf	Tetrafluoropropene	$C_3H_2F_4$	114.04	−29.4	95	160	769
R-290	Propane	C_3H_8	44.10	−42.1	97	203	973
R-410a	Puron®		72.6	−48.5	73	480	1878
R-513a	Opteon® XP10		108.4	−29.1	97	175	778
R-717	Ammonia	NH_3	17.03	−33.3	132.5	237	1167
R-718	Water	H_2O	18.015	100.0	374.0	–	4.3
R-744	Carbon Dioxide	CO_2	44.01	−78.4	31.0	2291	7213

Source Adapted from *2017 ASHRAE Handbook—Fundamentals*, Chapter 29

gas appears at a specific pressure, is used as the boiling temperature for refrigerant blends such as R-410a. Note that the evaporation pressures at −15 °C of the refrigerants listed are greater than atmospheric pressure except for water. The condenser pressures at 30 °C are approximately 700–1100 kPa, except for carbon dioxide, which has a much greater pressure.

Example 2.1   Performance Comparison of R-22 and Propane (R-290)
Since the HCFC refrigerant R-22 has a global warming potential of 1600, it is currently being phased out. One possible replacement is the refrigerant propane (R-290). For a condensing temperature of 40 °C, and evaporation temperature of 0 °C, compare the evaporator pressure P_1, the condenser pressure P_2, the specific compressor work w_c, the specific refrigeration q_l, and the coefficient of performance COP for a refrigeration system with an isentropic compressor efficiency η_c of 0.8. The thermodynamic cycle is shown schematically in Fig. 2.6.

Solution
The properties of refrigerants R-22 and Propane (R-290) are obtained from ASHRAE property tables.
R-22 Properties:
   Saturated liquid at +40 °C, $P_3 = P_2 = 1534$ kPa, $h_3 = h_4 = 249.6$ kJ/kg.
   Saturated vapor at 0 °C, $P_1 = 498$ kPa, $h_1 = 405.0$ kJ/kg, $s_1 = 1.7501$ kJ/kg-K.
R-290 Properties:
   Saturated liquid at +40 °C, $P_3 = P_2 = 1369$ kPa, $h_3 = h_4 = 307.1$ kJ/kg.
   Saturated vapor at 0 °C, $P_1 = 474$ kPa, $h_1 = 574.9$ kJ/kg, $s_1 = 2.3724$ kJ/kg-K.
Cycle analysis:
   R-22 isentropic compressor: $s_2 = s_1 = 1.7501$ kJ/kg-K, so $h_{2s} = 435$ kJ/kg:

$$w_c = (h_{2s} - h_1)/\eta_c = (435 - 405.0)/0.8 = 37.5 \, \text{kJ/kg}.$$

   R-290 isentropic compressor: $s_2 = s_1 = 2.3724$ kJ/kg-K, so $h_{2s} = 630$ kJ/kg:

Fig. 2.6 Temperature–entropy states for Example 2.1

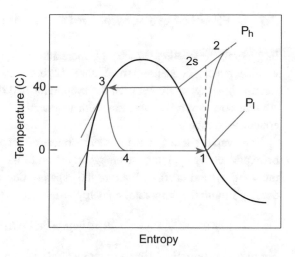

$$w_c = (h_{2s} - h_1)/\eta_c = (630 - 574.9)/0.8 = 68.8\,\text{kJ/kg}.$$

R-22 constant pressure evaporator:

$$q_l = h_1 - h_4 = 405.0 - 249.6 = 155.4\,\text{kJ/kg}.$$

R-290 constant pressure evaporator:

$$q_l = h_1 - h_4 = 574.9 - 307.1 = 267.8\,\text{kJ/kg}.$$

R-22 Coefficient of Performance:

$$COP = \frac{q_l}{w_c} = \frac{155.4}{37.5} = 4.14.$$

R-290 Coefficient of Performance:

$$COP = \frac{q_l}{w_c} = \frac{267.8}{68.8} = 3.89.$$

Discussion: The R-290 condenser pressure is 10% lower, and evaporator pressure is 5% lower, both comparable to R-22, so no high-pressure retrofit would be required. R-290 has almost double the specific compressor work w_c, and refrigeration q_l, due to its greater latent heat. The work and refrigeration effects balance, so the COP values are very close, with only a 6% decrease for the R-290 replacement. Note that there is a flammability issue with propane, since it is a safety class A3 refrigerant.

2.7   Effect of Evaporator and Condenser Temperature

In general, the cooling capacity $\dot{Q}_l$ increases with increasing evaporator temperature and with decreasing condenser temperature. Accordingly, the capacity is greatest when the difference in temperature between the evaporator and the condenser is minimized. Also, the compressor power $\dot{W}_c$ increases with increasing evaporator temperatures and increasing condenser temperatures.

The evaporator and condenser temperature effect on capacity and compressor power can be explained by applying the energy equation to the evaporator and the compressor. From the open system energy equation, the capacity depends on the product of the refrigerant density ρ_1 and the evaporator enthalpy difference $(h_1 - h_4)$:

$$\dot{Q}_l = \dot{m}q_l = \rho_1 \dot{V}_1(h_1 - h_4). \tag{2.16}$$

For an increase in the evaporator temperature as shown in Fig. 2.7, the evaporator pressure is increased, which will increase the vapor density at state 1. It will also decrease the quality of the refrigerant entering the evaporator at state 4, which increases the evaporator enthalpy difference $(h_1 - h_4)$. These two factors combine to increase the cooling capacity of the chiller.

As the condenser temperature is increased as shown in Fig. 2.8, the flow rate will decrease because of the higher discharge pressure P_2. The quality of the refrigerant entering the evaporator at state 4 will also increase, which will decrease the evaporator enthalpy difference $(h_1 - h_4)$. Both of these factors will decrease the cooling capacity.

Fig. 2.7 Effect of increase in evaporator temperature

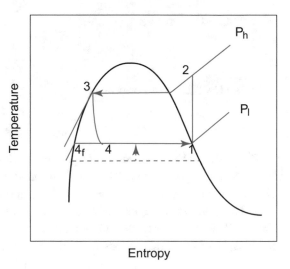

Fig. 2.8 Effect of increase in condenser temperature

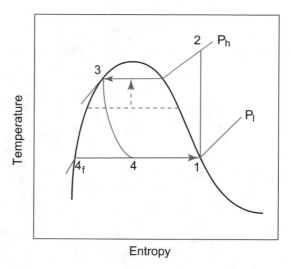

The compressor power is a function of the product of the refrigerant density ρ_1 and the compressor enthalpy change $(h_2 - h_1)$:

$$\dot{W}_c = \dot{m} w_c = \rho_1 \dot{V}_1 (h_2 - h_1). \tag{2.17}$$

For the compressor, as the evaporator temperature is increased at state 1, the mass flow rate is increased with the increase in inlet density, and the enthalpy change $h_2 - h_1$ decreases. The net effect is to increase the required compressor power due to the increase in mass flow. With an increase in condenser temperature, there is an increase in the enthalpy change $(h_2 - h_1)$, and a corresponding increase in the compressor power.

Cooling capacity and compressor power are presented in a representative compressor map in Fig. 2.9. The specific compressor represented in Fig. 2.9 is a 10 ton reciprocating four-cylinder semi-hermetic compressor operating at 1725 rpm with an air-to-liquid evaporator and condenser. The compressor performance is given as a function of the condensing and evaporating temperatures. The compressor map graphically indicates that to maximize capacity, a vapor compression system should be designed to operate at the highest practical evaporator temperature, and the lowest condensing temperature. Also, the cycle performance is more sensitive to changes in the evaporator temperature than changes in the condenser temperature.

For reasonable heat transfer rates, the temperature of the fluid (air or water) used to cool the condenser should be about $10\,^{\circ}\text{C}$ less than the condensing refrigerant, and the temperature of the heat transfer fluid used in the evaporator should be about $10\,^{\circ}\text{C}$ greater than the evaporating refrigerant.

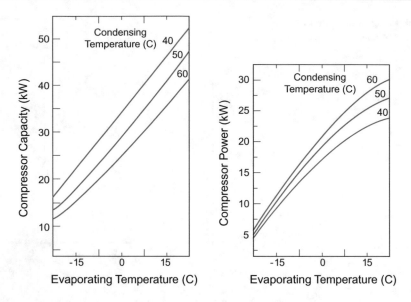

Fig. 2.9 Representative cooling capacity and compressor power map

Example 2.2   *Effect of Evaporator and Condenser Temperatures*

An ideal vapor compression system operates with R-134a as the working fluid. The temperature of the refrigerant in the evaporator is $-20\,°C$ and in the condenser is $40\,°C$.

(a) What is the refrigeration, q_l, and the coefficient of performance, COP?

(b) If the evaporator temperature is changed to $+5\,°C$, what is the effect on q_l and COP?

(c) If the condenser temperature is changed to $+50\,°C$, what is the effect on q_l and COP?

Solution

The T-s diagrams for the increased evaporator temperature and the increased condenser temperature are given in Figs. 2.7 and 2.8. The properties of refrigerant R-134a are obtained from property tables.

R-134a saturation properties:

Saturated vapor at $-20\,°C$, $h_1 = 386.1$ kJ/kg, $h_{4f} = 173.7$ kJ/kg, $P_1 = 133.7$ kPa, $s_1 = 1.739$ kJ/kg-K.

Saturated vapor at $+5\,°C$, $h_1 = 401.3$ kJ/kg, $h_{4f} = 206.75$ kJ/kg, $P_1 = 350.9$ kPa, $s_1 = 1.724$ kJ/kg-K.

Saturated liquid at $+40\,°C$, $P_3 = 1017$ kPa, $h_3 = 256.5$ kJ/kg.

Saturated liquid at $+50\,°C$, $P_3 = 1318$ kPa, $h_3 = 271.8$ kJ/kg.

(a) Baseline cycle analysis:

Isentropic compressor: $s_2 = s_1 = 1.739\,\text{kJ/kg-K}$, $P_2 = P_3 = 1017\,\text{kPa}$, $h_2 = 428.4$ kJ/kg:

$$w_c = h_2 - h_1 = 428.4 - 386.1 = 42.3\,\text{kJ/kg}.$$

Constant pressure condenser:

$$q_h = h_2 - h_3 = 428.4 - 256.5 = 171.9\,\text{kJ/kg}.$$

Expansion valve:

$$h_4 = h_3 = 256.5\,\text{kJ/kg}$$

$$x = \frac{h_4 - h_{4f}}{h_1 - h_{4f}} = \frac{265.5 - 173.7}{386.1 - 173.7} = 0.43.$$

Constant pressure evaporator:

$$q_l = h_1 - h_4 = 386.1 - 256.5 = 129.6\,\text{kJ/kg}.$$

Coefficient of Performance:

$$COP = \frac{q_l}{w_c} = \frac{129.6}{42.3} = 3.06.$$

(b) The cycle parameters for the three cases are tabulated in Table 2.4. As the evaporator temperature increases from -20 to $+5\,°C$, the refrigeration q_l and the COP increase, while the compression work w_c decreases. The increase in the refrigeration effect at higher evaporator temperatures is the result of the decrease in the quality of the working fluid entering the evaporator. This increases the amount of fluid that can be vaporized. The specific compressor work decreases since the pressure difference decreases.

(c) Conversely, as the condenser temperature increases from $+40$ to $+50\,°C$, the refrigeration q_l and the COP decrease, and the compression work w_c increases. The decrease in q_l and the COP is due to the increase in the quality of the working fluid entering the evaporator. The specific compressor work increases since the pressure difference increases.

Table 2.4 Results for Example 2.2

T_1 (C)	T_3 (C)	P_1 (kPa)	P_2 (kPa)	q_l (kJ/kg)	w_c (kJ/kg)	x (–)	COP (–)
−20	40	133.7	1017	129.6	42.3	0.43	3.06
+5	40	350.9	1017	144.8	22.2	0.26	6.52
−20	50	133.7	1318	114.3	49.1	0.46	2.33

2.8   Part Load Performance

The performance at part capacity, i.e., part load, is an important performance metric for cooling systems. Chillers are sized to meet a maximum design cooling load, but operate at this maximum load a small fraction of the time. Most of the time, the cooling system will operate at part load to meet loads that are less than the design load.

With centrifugal compressors, the load or capacity is controlled by varying the rotational speed of the compressor using variable frequency drives or adjusting the inlet guide vanes to change the inlet swirl of the working fluid. With multi-cylinder reciprocating compressors, the capacity is controlled by closing inlet valves and unloading cylinders so there is no refrigerant flow through a given cylinder. Small capacity systems will cycle the compressor off and on. However, due to refrigerant migration from the high-pressure condenser to the low-pressure evaporator when the unit is off, on–off cycling will reduce the part load efficiency.

We define the Part Load Ratio (PLR) of the cooling system as the ratio of the part load $\dot{Q}_l$ to the full load capacity $\dot{Q}_{full}$,

$$\text{PLR} = \frac{\dot{Q}_l}{\dot{Q}_{full}}. \tag{2.18}$$

The compressor power $\dot{W}_c$ required at part load is expressed in terms of PLR as

$$\dot{W}_c = \dot{W}_{full}\,(a + b\,PLR + c\,PLR^2). \tag{2.19}$$

The part load coefficients a, b, and c are found from manufacturers' data. Typical values are given in Table 2.5. The coefficients result from tests holding the entering condenser water temperature and leaving chilled water temperatures constant. Note that the sum of the constants $a + b + c \simeq 1$ to match the end condition $\dot{W}_c = \dot{W}_{full}$ at $PLR = 1$.

The coefficient of performance COP of the cooling system at part load is therefore

$$\begin{aligned}
\text{COP} &= \frac{\dot{Q}_l}{\dot{W}_c} \\
&= \frac{\dot{Q}_{full}\,PLR}{\dot{W}_{full}\,[a + b\,PLR + c\,PLR^2]}.
\end{aligned} \tag{2.20}$$

Table 2.5 Typical part load coefficients for compressors

Compressor type	a	b	c
Reciprocating	0.02	1.43	−0.48
Hermetic	0.16	0.32	0.52
Centrifugal	0.05	0.54	0.39

Example 2.3   Part Load Performance of Centrifugal Chiller
A cooling system with a centrifugal compressor has a rated COP = 3.6 and a full load capacity of 500 kW. Using the data in Table 2.5, what is the compressor power input W_c at a part load $\dot{Q}_l = 300$ kW?

Solution
The part load ratio PLR is

$$\text{PLR} = \frac{\dot{Q}_l}{\dot{Q}_{\text{full}}} = \frac{300}{500} = 0.6.$$

The full load compressor power is

$$\dot{W}_{\text{full}} = \frac{\dot{Q}_{\text{full}}}{COP} = \frac{500}{3.6} = 139.9\,\text{kW}.$$

The part load compressor power input is

$$\dot{W}_c = \dot{W}_{\text{full}}\,[a + b\,PLR + c\,PLR^2]$$
$$= (139.9)[0.049 + 0.545\,(0.6) + 0.389\,(0.6)^2]$$
$$= 72\,\text{kW}.$$

2.9   Multistage Vapor Compression Systems

When the design temperature difference between the evaporator and the condenser is larger than about 50 °C, multistage vapor compression with intercooling is used to keep the pressure ratio from exceeding values greater than five to one, prevent excessive condenser temperatures, and maintain acceptable volumetric efficiency. The vapor is compressed in stages, and cooled between each stage, reducing the overall compression work required and the compressor discharge temperature.

Two-stage systems are used with evaporator temperatures between −20° and −60 °C, and three-stage systems below −60 °C. For example, in frozen food industries the evaporator temperature is as low as −40 °C, requiring multistage compression. Both reciprocating and centrifugal compressors are used in multistage systems. A common intercooling technique in multistage compression is the use of a flash chamber, which is a device that separates the vapor and liquid phases of the refrigerant. The vapor from the flash chamber is used to cool the refrigerant exiting the low-pressure compressor.

An ideal two-stage vapor compression cycle with a flash chamber is shown on the T-s diagram in Fig. 2.10. The components of this vapor compression cycle are given in Fig. 2.11. The cycle operation is as follows. Saturated vapor at pressure P_l is compressed isentropically by the low-pressure compressor from state 1 to the intermediate pressure P_i at state 2. In the mixing chamber, the state 2 superheated vapor is mixed with the state 9 saturated vapor from the flash chamber, reducing the mixture temperature to state 3. The state 3 working

Fig. 2.10 Ideal two-stage
vapor compression cycle

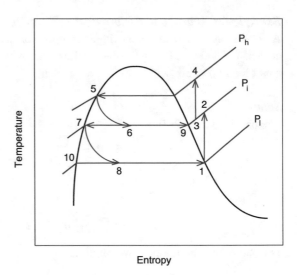

fluid mixture is compressed isentropically to state 4 at pressure P_h by the high-pressure
compressor. From state 4 to state 5, thermal energy is transferred from the working fluid to
the environment by condensation from a superheated vapor to a saturated liquid at constant
pressure P_h.

The working fluid then passes through the high-pressure throttling valve, expanding to
state 6 which is a two-phase state at pressure P_i, and enters the flash chamber. The flash
chamber separates the two-phase mixture into saturated vapor at state 9 and saturated liquid
at state 7. The state 9 saturated vapor fraction is mixed with the state 2 superheated vapor
in the mixing chamber to form the state 3 mixture. The state 7 saturated liquid fraction is
expanded through the low-pressure throttle valve to state 8, forming a two-phase mixture at
the evaporator pressure P_l. Finally, thermal energy, i.e., heat, is transferred from the body
to be cooled to the working fluid through an evaporation process from the two-phase region
at state 8 to a saturated vapor at state 1.

The multistage vapor compression cycle has two additional components relative to a
single-stage vapor compression cycle, the mixing chamber and the flash chamber. Thermodynamic analysis of this cycle requires the application of the open system energy equation
to these two components. The energy equation applied to the mixing chamber assuming
zero work and heat transfer is

$$\dot{m}_3 h_3 = \dot{m}_9 h_9 + \dot{m}_2 h_2$$

$$h_3 = \frac{\dot{m}_9}{\dot{m}_3} h_9 + \frac{\dot{m}_2}{\dot{m}_3} h_2. \tag{2.21}$$

The flash chamber separates the saturated vapor and saturated liquid with flow rate ratios
given by the quality x_6,

Fig. 2.11 Two-stage vapor compression cycle components

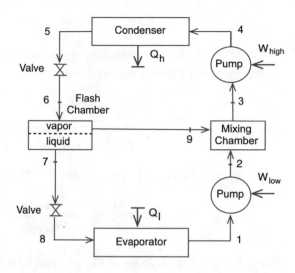

$$\frac{\dot{m}_9}{\dot{m}_3} = \frac{\dot{m}_9}{\dot{m}_6} = x_6$$

$$\frac{\dot{m}_2}{\dot{m}_3} = \frac{\dot{m}_7}{\dot{m}_6} = 1 - x_6. \tag{2.22}$$

The enthalpy h_3 can be expressed as

$$h_3 = x_6 h_9 + (1 - x_6)h_2. \tag{2.23}$$

The energy equation applied to the throttling valves in the top and bottom stages, assuming zero work and heat transfer yields

$$h_6 = h_5 = x_6 h_9 + (1 - x_6)h_7$$

$$x_6 = \frac{h_5 - h_7}{h_9 - h_7} \tag{2.24}$$

$$h_8 = h_7 = x_8 h_1 + (1 - x_8)h_{10}$$

$$x_8 = \frac{h_7 - h_{10}}{h_1 - h_{10}}. \tag{2.25}$$

With an ideal cycle, the low- and high-pressure compression are isentropic processes,

$$T_2 = T_1 \left(\frac{P_2}{P_1}\right)^{(\gamma-1)/\gamma}$$

$$T_4 = T_3 \left(\frac{P_4}{P_3}\right)^{(\gamma-1)/\gamma}. \tag{2.26}$$

The thermal energy transfer $\dot{Q}_l$ to the evaporator from the object to be cooled is

$$\dot{Q}_l = \dot{m}_7(h_1 - h_8)$$
$$= \dot{m}_6(1 - x_6)(h_1 - h_8). \tag{2.27}$$

The work $\dot{W}_t$ of the two compressors is

$$\dot{W}_t = \dot{W}_{low} + \dot{W}_{high}$$
$$= \dot{m}_7(h_2 - h_1) + \dot{m}_3(h_4 - h_3) \tag{2.28}$$
$$= \dot{m}_6(1 - x_6)(h_2 - h_1) + \dot{m}_6(h_4 - h_3).$$

Finally, the Coefficient of Performance (COP) of the two-stage cycle can be expressed as

$$\text{COP} = \frac{\dot{Q}_l}{\dot{W}_t} = \frac{(1 - x_6)(h_1 - h_8)}{(1 - x_6)(h_2 - h_1) + (h_4 - h_3)}. \tag{2.29}$$

Example 2.4   Two-Stage Vapor Compression with R-717 Ammonia

A two-stage vapor compression cycle uses ammonia as the working fluid. The cycle has an evaporator temperature of $-45\,°C$, a condenser temperature of $+38\,°C$, and an intermediate temperature of $-11\,°C$. (a) What are the pressure ratios of the two compressors? (b) What is the coefficient of performance (COP)? (c) What is the maximum cycle temperature? (d) Compare the two-stage cycle parameters with a single-stage cycle. Assume a constant specific heat ratio $\gamma = 1.38$ for ammonia.

Solution

Property data for saturated ammonia fluid and vapor is given in Table 2.6.

(a) The low-pressure compressor pressure ratio is

$$\frac{P_i}{P_l} = \frac{0.2789}{0.0545} = 5.1.$$

Table 2.6 Saturated ammonia property data for Example 2.4

State	Pressure P (MPa)	Temperature T (K)	Temperature T (°C)	Enthalpy h (kJ/kg)
1	0.0545	228	−45	1399.5
5	1.47	311	+38	381.0
7	0.279	263	−11	149.5
9	0.279	263	−11	1449.5
10	0.0545	228	−45	−2.78

The high-pressure compressor pressure ratio is

$$\frac{P_h}{P_i} = \frac{1.47}{0.279} = 5.3.$$

The single-stage compressor pressure ratio is

$$\frac{P_h}{P_l} = \frac{1.47}{0.0545} = 26.7.$$

Note that the two-stage compressors have pressure ratios of about 5, while the single-stage compressor has a high-pressure ratio of about 27.

(b) The top stage throttling valve exiting quality and enthalpy are

$$x_6 = \frac{h_5 - h_7}{h_9 - h_7} = \frac{381 - 149.5}{1449.5 - 149.5} = 0.178$$

$$h_6 = h_5 = 381 \text{ kJ/kg.}$$

The bottom stage throttling valve exiting quality and enthalpy are

$$x_8 = \frac{h_7 - h_{10}}{h_1 - h_{10}} = \frac{149.5 - (-2.78)}{1399.5 - (-2.78)} = 0.108$$

$$h_8 = h_7 = 149.5 \text{ kJ/kg.}$$

The low-pressure compressor exit temperature is

$$T_2 = T_1 \left(\frac{P_2}{P_1}\right)^{(\gamma-1)/\gamma}$$

$$= (228.15) \left(\frac{0.279}{0.0545}\right)^{(1.38-1)/1.38}$$

$$= 357.7 \text{ K} = 84.5 \,°\text{C.}$$

This temperature is in the superheated region. Given (P_2, T_2), from the superheat tables, $h_2 = 1650$ kJ/kg.

The outlet enthalpy h_3 of the mixing chamber is

$$h_3 = x_6 h_9 + (1 - x_6)h_2$$
$$= (0.178)(1449.5) + (1 - 0.178))1650 = 1614 \text{ kJ/kg.}$$

This enthalpy is in the superheated region. Given (P_3, h_3), from the superheat tables, $T_3 = 69\,°\text{C} = 342.15$ K. The mixing chamber reduces the working fluid temperature by about 15 C.

The high-pressure compressor exit temperature is

$$T_4 = T_3 \left(\frac{P_4}{P_3} \right)^{(\gamma-1)/\gamma}$$

$$= (342.15) \left(\frac{1.47}{0.279} \right)^{(1.38-1)/1.38}$$

$$= 540\,\mathrm{K} = 267\,^\circ\mathrm{C}.$$

This temperature is in the superheated region. Given (P_4, T_4), from the superheat tables, $h_4 = 2076\,\mathrm{kJ/kg}$.

The COP is

$$\mathrm{COP} = \frac{(1 - x_6)(h_1 - h_8)}{(1 - x_6)(h_2 - h_1) + (h_4 - h_3)}$$

$$= \frac{(1 - 0.178)(1399.5 - 148.7)}{(1 - 0.178)(1650 - 1399.5) + (2076 - 1614)}$$

$$= \frac{1028}{205.9 + 462} = 1.54.$$

(c) The maximum cycle temperature is $T_4 = 267\,^\circ\mathrm{C}$.

(d) With reference to the single compressor state points labeled in Fig. 2.3, the single compressor exit temperature T_2 is

$$T_2 = T_1 \left(\frac{P_2}{P_1} \right)^{(\gamma-1)/\gamma}$$

$$= (228.15) \left(\frac{1.47}{0.0545} \right)^{(1.38-1)/1.38}$$

$$= 564.6\,\mathrm{K} = 291.4\,^\circ\mathrm{C}.$$

The maximum single-stage cycle temperature is $291.4\,^\circ\mathrm{C}$, about $24\,^\circ\mathrm{C}$ greater than that of the two-stage cycle.

The temperature T_2 is in the superheated region. Given (P_2, T_2), from the superheat tables, $h_2 = 2138\,\mathrm{kJ/kg}$.

The enthalpy h_3 at the end of the heat rejection from state 2 to state 3 is the saturated liquid enthalpy h_f at $38\,^\circ\mathrm{C}$ given in Table 2.6. The enthalpy $h_4 = h_3$ since the enthalpy is unchanged from state 3 to state 4 across the throttle valve:

$$h_4 = h_3 = 381.0\,\mathrm{kJ/kg}.$$

The single-stage COP is

$$\mathrm{COP} = \frac{h_1 - h_4}{h_2 - h_1} = \frac{1399.5 - 381.0}{2138 - 1399.5} = 1.38.$$

This COP is about 10% less than the two-stage COP of 1.54.

2.10   Further Reading

Two thermodynamics textbooks that provide additional information about various vapor compression cooling cycles are Cengel (2014) and Sonntag (2020).

References

Cengel Y, Boles M (2014) Thermodynamics: An Engineering Approach. McGraw Hill, New York
Sonntag R, Borgnakke C (2020) Fundamentals of Thermodynamics. John Wiley, New York

Evaporative, Absorption, and Gas Cooling Cycles 3

3.1 Introduction

This chapter covers additional refrigeration and air conditioning methods, namely evaporative cooling, absorption refrigeration, and gas cooling. Evaporative cooling uses the latent heat of evaporation of water to cool an air stream. It is used in dry climates that have a significant difference between the dry and wet bulb air temperatures. However, the temperature decrease of the air stream in evaporative cooling is smaller when compared to vapor compression cooling. Absorption refrigeration systems are used where a source of low grade heat such as solar energy or waste heat is available as it uses thermal energy instead of mechanical energy as the driving energy input. Absorption systems function on the principle that liquid absorbents can attract and retain the vapor phase of a refrigerant at a relatively low temperature. The refrigerant is absorbed at low pressure and temperature in an absorber, and driven off as vapor in a high pressure and temperature generator. The gas refrigeration cycle is a reversed Brayton cycle, and operates in the superheated vapor regime, with no phase change of the working fluid. It is used in aircraft cabin cooling and to liquify air.

3.2 Evaporative Cooling

Evaporative cooling can be used in dry or low humidity climates for cooling purposes. These climates include the southwestern US, Mexico, Australia, and the Middle East. This technique takes advantage of the latent heat of vaporization of water. As a hot stream of air evaporates a water droplet or film, the water absorbs thermal energy from the air, and the air is cooled. The potential for evaporative cooling depends on the difference between the dry and wet bulb temperatures of the ambient air, i.e., on the initial air temperature and humidity.

A. T. Kirkpatrick, *Introduction to Refrigeration and Air Conditioning Systems*,
Synthesis Lectures on Mechanical Engineering,
https://doi.org/10.1007/978-3-031-16776-8_3

Evaporative cooling is much less expensive than vapor compression cooling, since no mechanical components are needed. It is less expensive to install and maintain, and has lower power consumption. The working fluid is water, not a halocarbon refrigerant. However, the temperature decrease of the air stream is smaller when compared to vapor compression cooling. There is no dehumidification in direct evaporative cooling, so evaporative cooling can result in high humidity levels in indoor spaces.

3.2.1   Direct Evaporative Cooling

There are two methods of evaporative cooling: direct and indirect. In direct evaporative cooling, as shown in Fig. 3.1, water comes into contact with the inlet air either as a spray or on a wetted porous matrix. With a wetted matrix, inlet air from the environment is drawn over the wetted pad and discharged into the space to be cooled. Wood fiber or plastic pads are used to maximize the surface contact area. A water pump with a make-up sump water supply is used to keep the porous matrix saturated. The pump supplies water to the top of the pad. As the water drains through the pad, a portion of the water evaporates, and the remainder flows back to the sump. Make up water is added to the sump at the same rate as the evaporation from the wetted pad.

The direct evaporative cooling process is modeled as an adiabatic saturation process. From the open system energy equation, if there is no work or heat transfer to the surroundings, then the enthalpy h and the wet bulb temperature T_{wb} of the air stream are essentially unchanged during the evaporation process. Accordingly, on the psychrometric chart, the process then

Fig. 3.1 Direct evaporative cooling

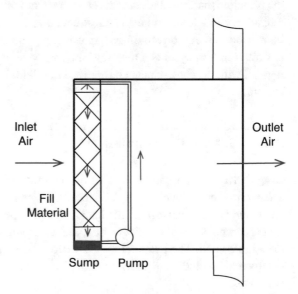

follows a line of constant wet bulb temperature, toward higher humidity values, until the air stream is saturated.

In practice, the air flow exiting an evaporative cooler is not fully saturated. The degree of saturation is defined by the effectiveness ϵ, a dimensionless measure of how close the exiting (e) dry bulb temperature T_{db}^e is to the inlet (i) wet bulb temperature T_{wb}^i.

$$\epsilon = \frac{T_{db}^i - T_{db}^e}{T_{db}^i - T_{wb}^i} \tag{3.1}$$

Example 3.1   Direct Evaporative Cooling

What is the dry bulb temperature, and the absolute and relative humidity values of air produced by a direct evaporative cooler with an effectiveness of 85% in a location in Arizona where the ambient environmental conditions are 40°C dry bulb temperature and 24 °C wet bulb temperature? What are the entering absolute and relative humidity values? Assume sea level conditions.

Solution

The evaporative cooling path is plotted on the psychrometric chart in Fig. 3.2. The wet bulb temperature of 24 °C is constant during the evaporative cooling from state 1 to state 2. From Fig. 3.2, the entering absolute humidity at state 1 is 0.0124 kg_v/kg_a, and relative humidity is about 27%.

Rearranging Eq. (3.1) for the exiting dry bulb temperature, we have

$$T_{db}^e = T_{db}^i - \epsilon\,(T_{db}^i - T_{wb}^i)$$
$$= 40 - 0.85(40 - 24) = 26.4°C$$

The exiting absolute humidity at state 2 for a dry bulb temperature of 26.4 °C is 0.0178 kg_v/kg_a, and the relative humidity is about 82%. Note the large increase in both absolute and relative humidity due to the evaporation of water into the air stream.

3.2.2   Indirect Evaporative Cooling

Indirect evaporative coolers are heat exchangers with a dry primary channel and a wet secondary channel separated by an impermeable wall. The basic configuration of a parallel flow indirect evaporative cooler is sketched in Fig. 3.3. The primary air stream is cooled sensibly by convective heat transfer to the impermeable wall without addition of moisture and is the cold air supply for a space. The process is not adiabatic since there is heat transfer from the hot air in the dry channel to the water film in the wet channel. As air passes through the secondary channel, the water in the wet media evaporates, cooling and humidifying the airstream, which is exhausted to the atmosphere. An advantage of indirect evaporative cooling is that there is no added humidity to the primary air stream.

Fig. 3.2 Evaporative cooling
path on psychrometric chart
(Example 3.2.1)

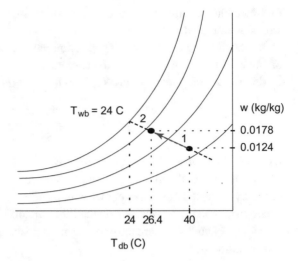

Fig. 3.3 Indirect evaporative
cooling

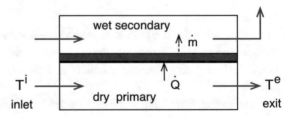

As discussed earlier, the ambient wet bulb temperature is the lower limit to the outlet
temperature of a direct evaporative cooler. In contrast, another advantage of indirect evaporative cooling is that the primary air stream can be cooled below it's wet bulb temperature,
since the wet bulb temperature of the secondary air stream becomes the lower limit of the
outlet temperature of the primary air stream. The effectiveness of an indirect evaporative
cooler is defined in the same manner as the effectiveness of a direct evaporative cooler, but
using $T_{db}^i - T_{wb}^i$, the difference between the primary air dry and wet bulb temperatures, as
the reference temperature difference.

$$\epsilon = \frac{T_{db}^i - T_{db}^e}{T_{db}^i - T_{wb}^i} \tag{3.2}$$

There are many configurations of indirect evaporative coolers. Most configurations incor-
porate some types of staging and regeneration. A single stage regenerative evaporative cooler
is shown in Fig. 3.4. In this counterflow configuration, the entering primary air is at temperature T_1. At the exit of the primary channel, a portion of the cooled exiting primary air at
temperature T_2 is routed back to the secondary channel. As the air flows along the secondary
channel, it is cooled and humidified by evaporation from the wet media in the secondary

Fig. 3.4 Single stage counter flow regenerative evaporative cooler

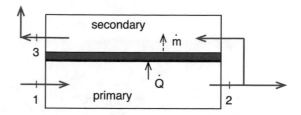

channel. At state 3, the exit of the secondary channel, the air stream is exhausted to the atmosphere. The temperature profiles in the primary and secondary channels are shown schematically in Fig. 3.5. The ambient air entering the primary channel is initially at temperature T_1. The exiting primary air is cooled to T_2, below the ambient wet bulb temperature. The secondary air initially decreases in temperature below T_2 due to evaporative cooling, then its temperature increases somewhat due to heat transfer from the liquid film, and it exits the secondary channel at temperature T_3,

This cooling process is also shown on the psychrometric diagram in Fig. 3.6. The temperature of the primary air decreases sensibly with a constant humidity ratio from T_1 to T_2. The secondary air, initially at T_2, is humidified to saturation at state 3. It is initially evaporatively cooled, then once saturated, is sensibly heated by the warmer liquid film in the wetted media in the secondary channel.

A two stage parallel flow indirect evaporative cooler is shown in Fig. 3.7. With this arrangement, a portion of the primary air exiting the dry first stage is branched to the wet channel of the second stage.

The process is sketched on the psychrometric diagram of Fig. 3.8. The primary air temperature decrease from state 1 to state 2 and then to state 3 is at constant humidity ratio. The secondary air flow follows a constant wet bulb path from state 1 to state 4 in stage 1 and

Fig. 3.5 Temperature profiles in a single stage evaporative cooler (Hasan 2010)

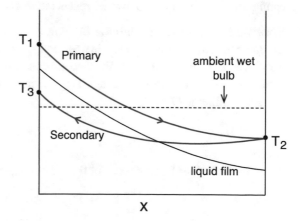

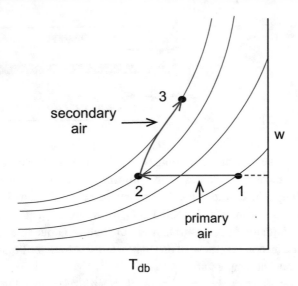

Fig. 3.6 Single stage regenerative cooling process

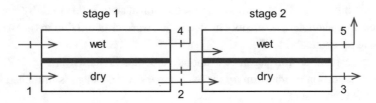

Fig. 3.7 Two stage parallel indirect evaporator cooler

from state 2 to state 5 in stage 2, and then is exhausted to the atmosphere. Other possible configurations are combined parallel-regeneration and two stage counterflow.

Example 3.2 Indirect Evaporative Cooling

Compare the effectiveness of a single stage counterflow regenerative chiller and a two stage parallel flow evaporative chiller. The ambient conditions are 30 °C dry bulb and 18.8 °C wet bulb. The primary air exiting the single stage chiller is 17 °C, and the primary air exiting the two stage chiller is 17.8 °C.

Solution

From Eq. (3.2), the effectiveness ϵ of the single stage chiller is

$$\epsilon = \frac{T_{db}^i - T_{db}^e}{T_{db}^i - T_{wb}^i} = \frac{30 - 17}{30 - 18.8} = 1.16$$

Fig. 3.8 Two stage
regenerative cooling process

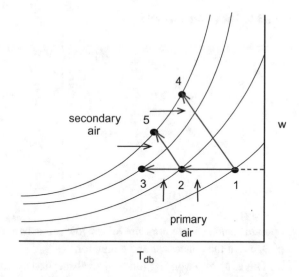

and the effectiveness ϵ of the two stage chiller is

$$\epsilon = \frac{T_{db}^i - T_{db}^e}{T_{db}^i - T_{wb}^i} = \frac{30 - 17.8}{30 - 18.8} = 1.09$$

Discussion: The effectiveness of indirect evaporative chillers can be greater than one, since the air is being cooled below the ambient air wet bulb temperature.

3.3   Absorption Refrigeration Cycles

Absorption refrigeration systems are used where a source of low grade heat such as solar energy or waste heat is available. The system components are more complex, thus more expensive, in an absorption system relative to a vapor compression system, so absorption systems are used mainly in large industrial and commercial building applications and in situations where electricity is not available, such as vacation homes, mobile homes and trailers. Absorption systems were the primary refrigeration technology used in the United States until replaced by lower cost vapor compression systems in the early to mid-1900s.

A schematic of a basic absorption cycle is shown in Fig. 3.9. A feature of this cycle is that thermal energy instead of mechanical energy is the driving energy input. Three components of the absorption cycle are the same as used in a vapor compression system, as absorption cycles also have a condenser, evaporator, and expansion valve. As shown in Fig. 3.9, an absorber and generator are used in place of a compressor, and the compression of the refrigerant occurs while it is in a liquid state, greatly reducing the pumping work. The cycle

Fig. 3.9 The absorption cycle

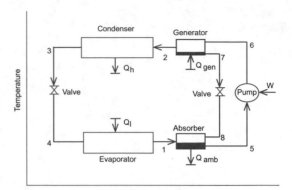

operates between two pressure levels: low pressure at the evaporator and absorber, and high pressure at the condenser and generator.

There are two working fluids in an absorption cycle, an absorbent, and a refrigerant. The absorption cycle uses the solubility of a refrigerant gas in an absorbent liquid to reduce the pumping energy required to compress the refrigerant. Absorption systems function on the principle that liquid absorbents can attract and retain the vapor phase of a refrigerant at a relatively low temperature, and that vapor solubility in an absorbent solution decreases as the solution temperature is increased. Absorption systems operate at sub-atmospheric or vacuum conditions. The refrigerant is absorbed at low pressure and temperature in the absorber, and driven off as vapor in the high pressure and temperature generator.

The two main absorbent/refrigerant combinations currently in use in absorption refrigeration cycles are lithium bromide/water ($LiBr/H_2O$) and water/ammonia (H_2O/NH_3). Water/ammonia mixtures are also called aqua ammonia or ammonium hydroxide. Other absorbents are lithium chloride and zinc chloride.

A water/ammonia absorption system is used in applications requiring refrigeration temperatures below $0\,°C$, as the freezing point of ammonia is $-78\,°C$. Both ammonia and water have a vapor pressure at standard operating temperatures. As thermal energy is added in the generator, both the ammonia and water will vaporize, with the boiling temperature depending on the local concentration. A vertical rectifying column is needed to separate the ammonia vapor from the water vapor. The water vapor has a higher saturation pressure than the ammonia, so most of the water vapor will condense and drain back to the generator, and the ammonia vapor will continue to rise through the rectifier without condensing.

Toxicity and flammability issues with ammonia require safeguards, adding to system cost and complexity. Ammonia can be very irritating to the eyes, throat, and breathing passages in small concentrations in air. The odor threshold for ammonia is between 5 - 50 parts per million (ppm) of air, and the permissible exposure limit (PEL) is 50 ppm averaged over 8 hours.

Since water is used as the refrigerant, LiBr systems are used only in applications requiring cooling at temperatures above $0\,°C$, i.e., air conditioning and process water cooling applications ranging from 4 to 10 °C. The vapor pressure of LiBr is very small, so the vapor produced in the generator is almost pure water, and so there are no absorbent/refrigerant separation issues that require a rectifier. Lithium bromide is nontoxic, nonflammable, chemically stable, somewhat corrosive to most common piping materials when exposed to air, and is also an irritant to skin and eyes.

3.3.1   Absorption Cycle Operation

The absorption cycle operation is shown in detail in Fig. 3.9. The refrigerant vapor leaving the evaporator (state 1) flows into the absorber and mixes with the absorbent liquid (state 8) sprayed at low temperature into the absorber. The absorption process is exothermic since the refrigerant vapor condenses into a liquid and goes into solution with the absorbent.

Heat transfer from the absorber to the ambient removes the latent heat and the heat of solution to maintain the low temperature of the absorber. The *strong* solution with a high concentration of refrigerant (state 5) is then pumped from the low pressure absorber to the higher generator pressure at state 6, and flows into the generator.

In the generator, the refrigerant/absorbent solution is heated by an external heat source which increases its temperature, causing the solubility of the refrigerant to decrease, so a portion of the refrigerant is driven out of the solution, leaving a *weak* solution with a higher concentration of absorbent. At state 2, the refrigerant vapor flows into the condenser, where it condenses to state 3, through transfer of thermal energy $\dot{Q}_{cond}$ to the ambient. The pressure of the refrigerant is reduced by the expansion valve, from state 3 to state 4. In the evaporator, there is heat transfer $\dot{Q}_{evap}$ to the refrigerant, which evaporates and returns to state 1. Meanwhile, the absorbent liquid in the generator at state 7 is throttled to the evaporator pressure in a separate flow loop with an expansion valve, and flows back to the absorber at state 8, completing the cycle.

The refrigerant and absorbent paths in the absorber and generator are plotted on the equilibrium pressure-temperature-concentration chart in Fig. 3.10. The solution temperature is plotted on the x-axis, and the system pressure is plotted on the y-axis. The three constant concentration lines are $X = 0$ (pure refrigerant), $X = X_{abs}$ (absorber concentration), and $X = X_{gen}$ (generator concentration), respectively. The paths follow alternating constant concentration and constant pressure lines. The refrigerant path is 1-5-6-2-1, and the absorbent path is 5-6-7-8-5. Notice that as the solution temperature is increased at constant pressure, the concentration of the refrigerant decreases due to the decreased solubility of the refrigerant in the absorbent, so $X_{gen} > X_{abs}$.

The LiBr concentration in a LiBr/water mixture depends on the mixture or solution temperature T_{mix} and the water saturation temperature T_{sat}. If saturation conditions are assumed, a curve-fit for the LiBr mass concentration valid for $0.50 < X < 0.65$ is given by Eq. (3.3) in units of kg_{LiBr}/kg_{mix}.

Fig. 3.10 Pressure-
Temperature-Concentration
diagram for absorption cycle

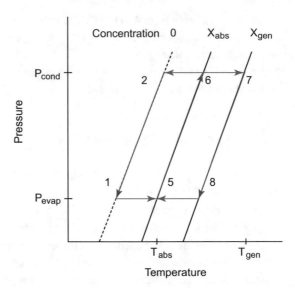

$$X = \frac{49.04 + 1.125\, T_{mix} - T_{sat}}{134.65 + 0.47\, T_{mix}} \tag{3.3}$$

3.3.2 Absorption Cycle Analysis

The working fluid of the absorption cycle has two components, the refrigerant and the
absorbent. Therefore, in addition to pressure and temperature, the analysis also needs to
account for the mass fraction or concentration X of the absorbent. In the analysis, Hasan
(2010), we first find the concentration and enthalpy at each point in the cycle for given pressures and temperatures, then apply the steady open system continuity and energy equations
to each of the cycle components to relate flowrate and heat transfer to enthalpy change.

Thermodynamic properties are needed for both the refrigerant and the absorbent. The
enthalpy of LiBr solutions as a function of concentration and temperature is given in Fig.
3.11. The enthalpy increases with increasing temperature and decreasing LiBr concentration.

The mass and energy balance equations for the generator are

$$\begin{aligned}\dot{m}_2 &= \dot{m}_6 - \dot{m}_7 \\ \dot{Q}_{gen} &= \dot{m}_2 h_2 + \dot{m}_7 h_7 - \dot{m}_6 h_6\end{aligned} \tag{3.4}$$

The mass and energy balance equations for the absorber are

$$\begin{aligned}\dot{m}_5 &= \dot{m}_1 + \dot{m}_8 \\ \dot{Q}_{abs} &= \dot{m}_1 h_1 + \dot{m}_8 h_8 - \dot{m}_5 h_5\end{aligned} \tag{3.5}$$

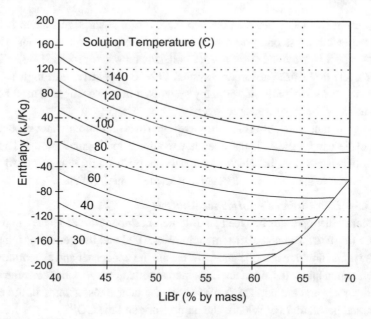

Fig. 3.11 Enthalpy-Temperature-Concentration diagram for H_2O-LiBr solution

The absorbent mass balance equation is

$$\dot{m}_5 X_5 = \dot{m}_7 X_7 \tag{3.6}$$

The energy balance equations for the condenser and evaporator with $\dot{m}_2 = \dot{m}_1$ are

$$\dot{Q}_{cond} = \dot{m}_2(h_2 - h_3)$$
$$\dot{Q}_{evap} = \dot{m}_1(h_1 - h_4) \tag{3.7}$$

The mass and energy balance equations for the valves and pump are

$$\dot{m}_3 = \dot{m}_4 \quad \dot{m}_7 = \dot{m}_8 \quad \dot{m}_6 = \dot{m}_5$$
$$h_3 = h_4 \quad h_7 = h_8 \tag{3.8}$$
$$\dot{W}_{pump} = \dot{m}_5 h_5 - \dot{m}_6 h_6$$

The coefficient of performance (COP) of an absorption system is somewhat different than a vapor compression system since the mechanical work $\dot{W}$ of liquid compression is small. The generator heat transfer $\dot{Q}_{gen}$ is the dominant term in the denominator of the COP definition:

$$COP = \frac{\dot{Q}_{evap}}{\dot{Q}_{gen} + \dot{W}} \approx \frac{\dot{Q}_{evap}}{\dot{Q}_{gen}} \tag{3.9}$$

The representative COP, between 0.6 and 0.75, for an absorption system is lower than that for a vapor compression system with the same temperature conditions. However, if the generator heat is provided from a solar collector or waste process heat, it is possible for the life cycle cost of the absorption system to be competitive with higher COP vapor compression systems. In addition, for many systems, a heat exchanger is used between the absorber and the generator for heat recovery.

By adding a non-condensible gas such as hydrogen to an ammonia-water vapor absorption system, a cycle can be created that does not require a mechanical pump. The third gas produces a two-phase flow in which vapor bubbles push liquid slugs upward in a small diameter tube above the generator, located at a low level in the cycle.

Example 3.3   LiBr Chiller Performance Analysis

A LiBr/water chiller operates between an evaporator temperature of $10\,^\circ$C and a condenser temperature of $40\,^\circ$C. The absorber temperature is $30\,^\circ$C and the generator temperature is $100\,^\circ$C. The pump flowrate is 0.60 kg/s. (a) What are the condenser and evaporator pressures and inlet/exit enthalpies? (b) What are the generator and absorber LiBr concentrations, and inlet/exit flowrates and enthalpies? (c) What are the heat transfer rates in the evaporator, condenser, and generator? (d) What is the pump work and the COP?

Solution

Assume that the exit conditions in the condenser and evaporator are saturated. If we neglect pressure drops in the system piping, then the generator will operate at the same pressure as the condenser, and the absorber will operate at the same pressure as the evaporator.

(a) From steam tables, at a condenser temperature of $40\,^\circ$C, the saturation pressure is 7.4 kPa, and the saturated liquid enthalpy $h_3 = h_4 = 167.5$ kJ/kg. At an evaporator temperature of $10\,^\circ$C, the saturation pressure is 1.2 kPa, and the saturated vapor enthalpy $h_1 = 2519$ kJ/kg. The water vapor leaves the generator at $100\,^\circ$C and 7.4 kPa, a superheated state with enthalpy $h_2 = 2674$ kJ/kg.

(b) From Eq. (3.3), the absorber LiBr concentration is

$$X = \frac{49.04 + 1.125\,T_{mix} - T_{sat}}{134.65 + 0.47\,T_{mix}} = \frac{49.04 + 1.125\,(30) - 10}{134.65 + 0.47\,(30)} = 0.49$$

and from Fig. 3.11 or the ASHRAE LiBr Tables, ASHRAE (2017), the enthalpy $h_5 = -165$ kJ/kg.

The generator LiBr concentration is

$$X = \frac{49.04 + 1.125\,T_{mix} - T_{sat}}{134.65 + 0.47\,T_{mix}} = \frac{49.04 + 1.125\,(100) - 40}{134.65 + 0.47\,(100)} = 0.67$$

and from Fig. 3.11 or the ASHRAE LiBr Tables, the enthalpy $h_7 = -55.8$ kJ/kg.

The LiBr mass flowrate from the generator to the absorber is

$$\dot{m}_7 = \dot{m}_5 \frac{X_5}{X_7} = (0.60)\frac{0.49}{0.67} = 0.44 \,\text{kg/s}$$

The water mass flowrate through the condenser and evaporator is thus

$$\dot{m}_2 = \dot{m}_6 - \dot{m}_7 = 0.60 - 0.44 = 0.16 \,\text{kg/s}$$

c.) The heat transfer rates of the evaporator and the condenser are

$$\dot{Q}_{evap} = \dot{m}_1(h_1 - h_4) = (0.16)(2519 - 167.5) = 376 \,\text{kW}$$
$$\dot{Q}_{cond} = \dot{m}_2(h_2 - h_3) = (0.16)(2674 - 167.5) = 401 \,\text{kW}$$

The heat transfer rate of the generator is

$$\dot{Q}_{gen} = \dot{m}_2 h_2 + \dot{m}_7 h_7 - \dot{m}_6 h_6$$
$$= (.16)(2674) + (0.44)(-55.8) - (0.60)(-165) = 502 \,\text{kW}$$

d.) From the ASHRAE LiBr Tables the specific volume of the LiBr solution in the absorber is $v = 6.58 \times 10^{-4} \text{m}^3/\text{kg}$. The specific pump work w_{pump} is therefore

$$w_{pump} = v(P_6 - P_5) = 6.58 \times 10^{-4}(7.4 - 1.2) = 4.08 \times 10^{-3} \,\text{kJ/kg}$$

$$\dot{W} = \dot{m} w_{pump} = (0.6)4.08 \times 10^{-3} = 2.4 \times 10^{-3} \,\text{kW}$$

Since $w_{pump} << h_5$, then $h_6 \simeq h_5$. The system COP is

$$COP = \frac{\dot{Q}_{evap}}{\dot{Q}_{gen}} = \frac{376}{502} = 0.75 \tag{3.10}$$

Discussion: Note that the pump power is much less than the heat transfer rate of the generator.

3.4   Gas Refrigeration Cycle

The gas refrigeration cycle is a reversed Brayton cycle, named for George Brayton (1830–1892), an American mechanical engineer. In contrast to the reversed vapor compression cycle, it operates entirely in the superheated vapor regime, with no change of phase of the working fluid. The T-s diagram for the gas refrigeration cycle is shown in Fig. 3.12. A major advantage of the cycle is that the working fluid, air (R-729), is non-toxic and freely available. However, gas refrigeration systems have much lower COP values than comparably sized vapor compression equipment, since the heat absorption and rejection is not isothermal.

The four basic processes in the ideal cycle are

Fig. 3.12 The gas refrigeration
cycle

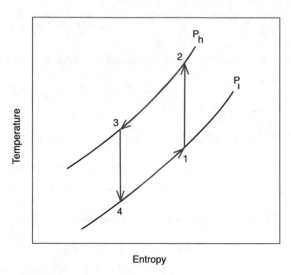

1 to 2   isentropic gas compression in compressor from P_l to P_h
2 to 3   constant pressure heat rejection to the environment
3 to 4   isentropic gas expansion in turbine from P_h to P_l
4 to 1   constant pressure heat absorption from object or space to be cooled

Since it is relatively light weight and compact, this cycle is used in aircraft cooling
systems. It is also used in cryogenic systems to liquify air and other gases. In the aircraft
application, the compressed air is drawn from the main gas turbine compressor, cooled, and
expanded to pressure P_l in a turbine, and then delivered to the aircraft cabin.

3.4.1   Thermodynamic Analysis

Using the open system energy equation and assuming isentropic compression, the compressor work and temperature change for the compression from 1 to 2,

$$w_c = h_2 - h_1 \tag{3.11}$$

$$\frac{T_2}{T_1} = \left(\frac{P_h}{P_l}\right)^{(\gamma-1)/\gamma} \tag{3.12}$$

The heat transfer to the environment is

$$q_h = h_3 - h_2 \tag{3.13}$$

The turbine work and temperature change for the isentropic expansion from 3-4 are

$$w_t = h_4 - h_3 \tag{3.14}$$

$$\frac{T_4}{T_3} = \left(\frac{P_l}{P_h}\right)^{(\gamma-1)/\gamma}$$

(3.15)

The heat transfer absorption from the object/space is

$$q_l = h_1 - h_4$$

(3.16)

The coefficient of performance COP, for constant specific heat, is

$$
\begin{aligned}
COP &= \frac{q_l}{w_{net}}\\
&= \frac{h_1 - h_4}{(h_2 - h_1) - (h_3 - h_4)}\\
&= \frac{T_1 - T_4}{(T_2 - T_3) - (T_1 - T_4)}\\
&= \frac{T_4(T_1/T_4 - 1)}{T_3(T_2/T_3 - 1) - T_4(T_1/T_4 - 1)}
\end{aligned}
$$

(3.17)

Since the compression and expansion occur between the same pressures, the temperature ratios are

$$\frac{T_1}{T_4} = \frac{T_2}{T_3}$$

(3.18)

and the COP can be expressed as a function of the pressure ratio:

$$
\begin{aligned}
COP &= \frac{T_4}{T_3 - T_4}\\
&= \left[\frac{T_3}{T_4} - 1\right]^{-1}\\
&= \left[\left(\frac{P_h}{P_l}\right)^{(\gamma-1)/\gamma} - 1\right]^{-1}
\end{aligned}
$$

(3.19)

Example 3.4   Gas Refrigeration COP

A gas refrigeration cycle operates with a low pressure $P_l = 350$ kPa and a high pressure $P_h = 1017$ kPa. What is the ideal cycle COP? Assume $\gamma = 1.4$ and compare with Example 2.1.

Solution

$$
\begin{aligned}
COP &= \left[\left(\frac{P_h}{P_l}\right)^{(\gamma-1)/\gamma} - 1\right]^{-1}\\
&= \left[\left(\frac{1017}{350}\right)^{(1.4-1)/1.4} - 1\right]^{-1}\\
&= 2.80
\end{aligned}
$$

Discussion: As indicated by Example 2.1, the COP for the same pressure difference with a vapor compression cycle is 6.52. The COP of gas compression cycles is always lower when compared to vapor compression cycles since the heat transfer processes are not isothermal.

3.5　Further Reading

Additional information about the modeling and performance of evaporative, absorption, and gas cooling cycles is given in the books by Kuehn et al. (1998) and Herold et al. (1996).

References

ASHRAE, (2017) ASHRAE Handbook – Fundamentals, Chapter 30 –Thermophysical Properties of Refrigerants. ASHRAE, Atlanta, GA

Hasan A (2010) Indirect evaporative cooling of air to a sub-wet bulb temperature. Applied Thermal Engineering 30:2460–2468

Herold K, Radermacher R, Klein S (1996) Absorption Chillers and Heat Pumps. CRC Press, Boca Raton, Florida

Kuehn T, Ramsey J, Threlkeld J (1998) Thermal Environmental Engineering. Prentice Hall, Hoboken, New Jersey

Fluid Flow in Refrigeration and Air Conditioning Systems

4

4.1 Introduction

The subject of this chapter is the fluid flow in refrigeration and air conditioning systems. Proper selection of pumps and fans in refrigeration and air conditioning systems requires knowledge of the fluid flow rate and associated pressure losses. In this chapter, we review the thermodynamic properties of fluids, develop the continuity and Bernoulli equations for fluid flow, and determine the fluid friction and resulting pressure drop for a given flow rate. The Bernoulli equation relates the flow velocity to the fluid pressure and changes in elevation. The pump and fan characteristics needed to produce a given flow rate and pressure increase are also presented.

The fluid flow coverage in the chapter begins with liquid flow in pipes and finishes with a discussion of air flow in ducts. In buildings, a Heating, Ventilating, and Air Conditioning (HVAC) system with supply and exhaust fans is used to supply conditioned air to rooms or zones, and to return exhaust air back to the outside environment. The supply air flow rate and temperature are chosen to maintain acceptable thermal comfort and air quality in the occupied zone. The supply fan delivers this conditioned air to the occupied zone to meet the cooling (or heating) load. An exhaust fan is used to remove hot and humid air from the occupied zone.

4.2 Fluid Properties

Four fluid properties useful for the calculation of fluid flow in refrigeration and air conditioning systems are pressure, density, viscosity, and thermal conductivity. Fluid pressure P is defined as the normal stress, i.e., the force per unit area normal to a surface at any point in a fluid at rest:

© The Author(s), under exclusive license to Springer Nature Switzerland AG 2023
A. T. Kirkpatrick, *Introduction to Refrigeration and Air Conditioning Systems*,
Synthesis Lectures on Mechanical Engineering,
https://doi.org/10.1007/978-3-031-16776-8_4

$$P = \lim_{A \to 0} \frac{F}{A}. \tag{4.1}$$

A variety of units are used for defining fluid pressure. In the SI system, the units of pressure are Pascals, where 1 Pa = 1 Newton/m^2. One standard atmosphere (1 atm) of pressure is defined as 101,325 Pa = 0.101 MPa = 14.696 lb/in^2. In a flowing fluid in a duct or pipe, pressures are commonly measured with a U-tube partially filled with water. One leg is open to the duct flow, and the other leg is open to the atmosphere, creating different water levels on either side of the tube.

As a consequence, duct static pressures are commonly expressed in millimeters (or inches) of water, resulting from the difference in the water levels. The hydrostatic equation, $P = \rho g h$, is used to convert the length unit to Pa, where ρ is the density of the fluid in the U-tube. For example, if a U-tube using water indicates a difference in the water levels of 1 cm (0.01 m), the static pressure relative to the atmosphere is $P = \rho g h = (998)(9.81)(0.01) = 98$ Pa.

Fluid pressures that are measured relative to the local atmospheric pressure are defined as gauge pressures. As indicated by Eq. (4.2), the absolute pressure is the sum of the gauge pressure and the local atmospheric pressure,

$$P_{abs} = P_{gauge} + P_{atm}. \tag{4.2}$$

The pressure in a liquid piping system should be above the vapor pressure to prevent vapor flashing and two-phase flow. In an air duct system, the absolute pressure should be maintained above atmospheric pressure to prevent air leakage into the system.

Fluid density ρ is defined as the fluid mass per unit volume. At 20°C, the density of water ρ_w is 998.2 kg/m^3. At the pressures typical of HVAC systems, water is assumed to be incompressible, and the density of water is commonly approximated as 1000 kg/m^3. The density of air depends on the temperature and pressure through the ideal gas equation,

$$\rho = \frac{P}{RT}. \tag{4.3}$$

At 20°C and one atmosphere pressure, the density of air is

$$\rho = \frac{P}{RT} = \frac{101,325}{(287)(273.15 + 20)} = 1.204 \, \text{kg/m}^3. \tag{4.4}$$

A related parameter is the specific volume $v = 1/\rho$, which is the reciprocal of density.

The two main transport properties of fluids are the dynamic viscosity μ with units of (Pa-s), and the thermal conductivity k with units of (W/m-K). In a fluid, these properties are related since both are functions of small-scale molecular motion. The property viscosity governs the transfer of momentum, and the property thermal conductivity governs the transfer of energy. The thermal conductivity of fluids is discussed in more detail in Chap. 5.

Viscosity is the property that describes the response of a fluid to an imposed shear stress, resulting in a velocity gradient and friction in the flow field. The dynamic viscosity μ is defined as the ratio of the shear stress to the velocity gradient,

$$\tau = \mu \frac{du}{dy}.$$

(4.5)

The viscosity of a fluid depends to a large extent on its temperature. For air, experiments indicate the dynamic viscosity increases with temperature approximately to 2/3 power:

$$\mu \sim T^{0.67}.$$

(4.6)

In 1893, W. Sutherland (1859–1911), an Australian physicist, developed an equation for the temperature dependence of the dynamic viscosity and thermal conductivity of gases known as Sutherland's Law. The equation is based on the kinetic theory of gases, which assumes the molecules in a gas are hard spheres in constant random motion colliding with each other and the surrounding walls. If the mean free path between collisions is assumed to be much larger than the distance between molecules, then the dynamic viscosity and thermal conductivity can be shown by kinetic theory to be proportional to the gas density, mean free path, and mean speed. However, since the mean free path is inversely proportional to the density, the dynamic viscosity and thermal conductivity are predicted to be independent of the gas pressure. The two-coefficient form of the Sutherland equation with temperature in degrees Kelvin is

$$\mu = \frac{a_1 \, T^{3/2}}{a_2 + T}$$

$$k = \frac{b_1 \, T^{3/2}}{b_2 + T}.$$

(4.7)

For air, the curve-fit coefficients in Sutherland equations are $a_1 = 1.512 \times 10^{-6}$ Pa-s/K$^{1/2}$, $a_2 = 120.0$ K, and $b_1 = 2.501 \times 10^{-3}$ W/m-K/K$^{1/2}$, $b_2 = 194.4$ K. At $T = 293$ K (20 °C),

$$\mu = \frac{(1.512 \times 10^6) \, 293^{3/2}}{120 + 293} = 1.836 \times 10^5 \text{ Pa-s}$$

$$k = \frac{(2.501 \times 10^3) \, 293^{3/2}}{194.4 + 293} = 2.573 \times 10^2 \text{ W/m-K}.$$

(4.8)

For water, the dynamic viscosity decreases with increasing temperature. A dynamic viscosity–temperature curve-fit (Perry 1999), with temperature in degrees Centigrade, is

$$\mu = \frac{0.1}{2.1482[D + (8078.4 + D^2)^{1/2}] - 120}$$

(4.9)

where $D = (T - 8.435)$ °C. At 20 °C, the dynamic viscosity of water is 1.0×10^3 Pa-s.

The kinematic viscosity v is defined as the dynamic viscosity μ divided by the fluid density ρ,

$$v = \frac{\mu}{\rho}.$$

(4.10)

The units of kinematic viscosity are m^2/s. At $20\,°C$ and standard atmospheric pressure, the kinematic viscosity of air is 1.52×10^5 m^2/s, and the kinematic viscosity of water is 1.0×10^6 m^2/s.

If ethylene or propylene glycol is used in a water system to prevent freezing, the resulting mixture will have a greater viscosity relative to water alone. In addition, the water–glycol mixture will have lower thermal conductivity and specific heat. More glycol solution must be circulated through a given coil to achieve the same heat transfer rates as with water, leading to higher fluid pressure drops, of the order of 1 m. Additional heat transfer surface is needed, leading to higher air-side pressure drops.

4.3   Fluid Flow Measurement

Flow measurements in pipes and ducts in buildings are needed for HVAC control and energy management. Devices have been developed to produce a pressure drop across the device which is then converted to velocity using a Bernoulli equation. Some of the various types of fluid flow meters that can be used include the following:

ASME Orifice It is often employed as a secondary standard to calibrate other meters; flow rate depends on the square root of the pressure drop across the orifice, so a range of orifice sizes is used to cover the air flow rate range.

Laminar Flow Meter A bundle of tubes (not necessarily round in cross section) sized so that the Reynolds number in each is well within the laminar regime; flow rate depends linearly on the pressure drop across the meter.

Critical Flow Nozzle A Venturi in which the flow is choked; the flow rate is then linearly dependent upon the delivery pressure (an external compressor is thus required) and independent of the pressure in the surge tank.

Turbine Meter The air flow rate is linearly dependent upon the rotational speed of the turbine.

Hot Wire Meter A hot wire anemometer is inserted into the flow to measure the centerline velocity; the air flow rate is proportional to centerline velocity.

No matter which of the various methods is used, measurements of temperature and pressure also have to be made. Key components of systems employed using these various meters are identified in Fig. 4.1. The calibration coefficients of the meters are a function of the Reynolds number of the flow through the meters. Correction factors are used to adjust measured data to standard atmospheric temperature and pressure conditions. The specific correction procedures are usually included in the meter operation manual.

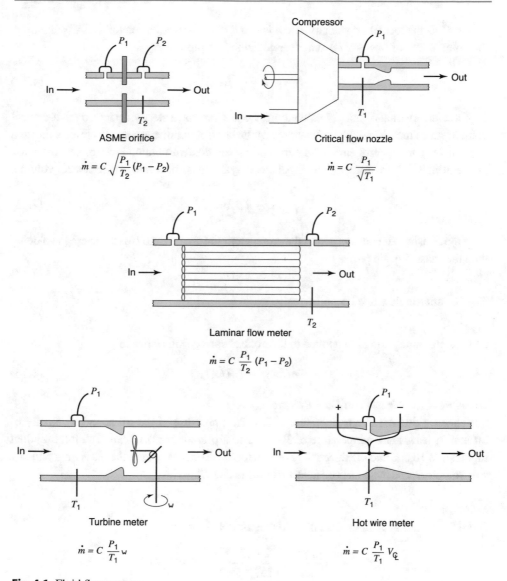

Fig. 4.1 Fluid flow meters

4.4 Continuity Equation

The mass flux (kg/s) through a pipe or duct with cross-sectional area A at a given location is a function of the flow density and velocity,

$$\dot{m} = \int_A \rho \vec{U} \cdot \hat{n} \, dA \tag{4.11}$$

where $\vec{U}$ is the local fluid velocity, and $\hat{n}$ is a unit outward normal vector. Correspondingly, the volume rate of flow (m³/s) at a given location in a pipe is

$$\dot{V} = \int_A \vec{U} \cdot \hat{n} \, dA. \tag{4.12}$$

The conservation of mass, i.e., continuity equation, for a steady flow through a control volume states that the net flux of mass out of the control volume is zero. Therefore, the area integral over the control surface A surrounding the control volume is zero, i.e., the mass flow exiting (e) the control volume is balanced by the mass flow into (i) the control volume,

$$\int_A \rho \vec{U} \cdot \hat{n} \, dA = 0. \tag{4.13}$$

If the density ρ is constant across the cross-sectional area A, and U is the average velocity, then the mass flux simplifies to

$$\dot{m} = \rho A U. \tag{4.14}$$

The volumetric flow rate becomes

$$\dot{V} = AU \tag{4.15}$$

and the continuity equation applied to the control surface simplifies to

$$(\rho U A)_i = (\rho U A)_e. \tag{4.16}$$

Example 4.1   Air Flow in Duct Systems
Air enters a diffuser at a temperature of 350 K, a pressure of 0.2 MPa, and a velocity of 30 m/s. It leaves at a temperature of 300 K, and a pressure of 0.1 MPa. The inlet (i) duct diameter is 10 cm, and the duct exit (e) diameter is 100 cm. What are the mass flow rate and exit velocity? The ideal gas constant R for air is 287 J/kg-K.

Solution
From the equation of state, the inlet and exit air densities are

$$\rho_i = \frac{P_i}{RT_i} = \frac{200,000}{(287)(350)} = 1.99 \, \text{kg/m}^3$$

$$\rho_e = \frac{P_e}{RT_e} = \frac{100,000}{(287)(300)} = 1.16 \, \text{kg/m}^3.$$

The mass flow rate is

$$\dot{m} = (\rho A U)_i$$

$$= (1.99) \frac{\pi}{4} (0.1)^2 (30) = 0.47 \, \text{kg/s}.$$

The continuity equation is

$$(\rho U A)_i = (\rho U A)_e .$$

Solving for the exit velocity,

$$\begin{aligned} U_e &= \frac{(\rho U A)_i}{(\rho A)_e} \\ &= \frac{(1.99)(30)(0.1)^2}{(1.16)(1.0)^2} = 0.51 \text{ m/s.} \end{aligned}$$

Note that the diffuser produces a reduction in both the density and the velocity.

4.5 Bernoulli's Equation

Bernoulli's equation relates the flow velocity in a piping system to the fluid pressure and elevation. It is named after Daniel Bernoulli (1700–1782), a Swiss scientist who made many contributions to science in the 1700s. Bernoulli's equation is based on the energy equation, which for a steady state control volume on a per unit mass basis (kJ/kg) is

$$q - w = \left(h + \frac{U^2}{2} + gz \right)_e - \left(h + \frac{U^2}{2} + gz \right)_i \tag{4.17}$$

where e denotes the fluid exiting the control volume, and i denotes the fluid entering the control volume. The second law of thermodynamics with temperature T at the inlet and exit control volume boundaries can be expressed as

$$q = T(s_e - s_i). \tag{4.18}$$

For a non-isentropic process, the integrated Gibbs equation from Chapter One is

$$T(s_e - s_i) = h_e - h_i - \int_i^e v dP. \tag{4.19}$$

If the work term w is zero, then upon substitution of Eqs. (4.18) and (4.19) into Eq. (4.17),

$$h_e - h_i - \int_i^e v dP = \left(h + \frac{U^2}{2} + gz \right)_e - \left(h + \frac{U^2}{2} + gz \right)_i . \tag{4.20}$$

Using an averaged specific volume $\bar{v}$ to simplify the pressure integral, and canceling common h terms,

$$0 = \left(\frac{U^2}{2} + gz \right)_e - \left(\frac{U^2}{2} + gz \right)_i + \bar{v}(P_e - P_i). \tag{4.21}$$

Upon substitution of the average fluid density $\rho = \bar{\rho} = 1/\bar{v}$ into Eq. (4.21), we have the Bernoulli equation for incompressible isothermal flow,

$$\left(\frac{P}{\rho} + \frac{U^2}{2} + gz\right)_i = \left(\frac{P}{\rho} + \frac{U^2}{2} + gz\right)_e. \tag{4.22}$$

The total pressure P_t in a fluid flow is defined as the sum of the static pressure P, and the velocity pressure $\rho U^2/2$,

$$P_t = P + \rho\frac{U^2}{2}. \tag{4.23}$$

The Bernoulli equation can be generalized for the case of flow with friction, i.e., mechanical energy loss, by adding a friction head H_f, defined relative to the kinetic energy per unit mass, $U^2/2$, with units of length,

$$H_f = K\frac{U^2}{2g}. \tag{4.24}$$

In cases where the work per unit mass term w is non-zero, a pump head H_p is added to the Bernoulli equation. It is defined relative to the pressure term and also has units of length,

$$H_p = \frac{\Delta P}{\rho g} = -\frac{\dot{W}}{\dot{m}g}. \tag{4.25}$$

Dividing Eq. (4.22) by g, we arrive at the generalized Bernoulli equation in head, i.e., length units,

$$\left(\frac{P}{\rho g} + \frac{U^2}{2g} + z\right)_i + H_p = \left(\frac{P}{\rho g} + \frac{U^2}{2g} + z\right)_e + H_f. \tag{4.26}$$

The pressure term is called the pressure head, and the velocity term is called the velocity head. Solving for the pump head H_p,

$$H_p = \left(\frac{P}{\rho g} + \frac{U^2}{2g} + z\right)_e - \left(\frac{P}{\rho g} + \frac{U^2}{2g} + z\right)_i + H_f. \tag{4.27}$$

Example 4.2   Pump Head in Piping Systems

A piping system delivers cooling water from a basement chiller to the roof of an office complex. The roof is 50 m above the basement level. The friction head loss H_f in the piping, valve, and fittings is 4 m. The water enters the pump at a static pressure head of 3 m, and is delivered to the roof tank at atmospheric pressure. The velocity at the pump suction is 1 m/s, and at the piping exit is 3 m/s. What is the required pump head?

Solution

The required pump head is

$$H_p = \left(\frac{P}{\rho g}\right)_e - \left(\frac{P}{\rho g}\right)_i + \frac{U_e^2 - U_i^2}{2g} + z_e - z_i + H_f$$

$$= 0 - 3 + \frac{3^2 - 1^2}{2(9.81)} + 50 - 0 + 4$$

$$= 51.4\,\text{m}.$$

 Note that the major component of the pump head is the elevation change from the basement
to the roof.

4.6   Energy Losses in Fluid Flow

The mechanical energy losses in fluid flow systems are due both to pipe or duct friction and
pressure losses in the fittings. The friction is modeled with a Darcy–Weisbach friction factor
f which relates the friction pressure loss ΔP (Pa) to the fluid kinetic energy per unit mass,

$$\Delta P = f \left(\frac{L}{D_h} \right) \frac{\rho U^2}{2} \tag{4.28}$$

where L is a length scale (m), D_h is the hydraulic diameter (m), ρ the fluid density (kg/m^3),
and U the average fluid velocity (m/s).

   The hydraulic diameter D_h is defined as the ratio of the duct cross-sectional area A_c to
the duct perimeter p,

$$D_h = \frac{4 A_c}{p}. \tag{4.29}$$

The friction factor depends on the pipe or duct roughness, ϵ, and the non-dimensional flow
Reynolds number Re,

$$Re = \frac{U D_h}{\nu} = \frac{4 \dot{m}}{\pi D_h \mu}. \tag{4.30}$$

The Reynolds number is named after Osborne Reynolds (1842–1912), an English engineering professor who proposed it in 1883 to characterize the condition in which the flow of
fluid in pipes transitions from laminar flow to turbulent flow. It is the ratio of the inertial to
viscous forces in the flow. The flow in pipes or ducts is laminar at Reynolds numbers below
about 2300, and becomes turbulent at Reynolds numbers above that value.

   A widely used formula, Eq. (4.31) (White 2015), for fluid friction in turbulent flow
($Re > 2300$) is

$$\frac{1}{f^{1/2}} = 0.8687 \left[\ln \left(\frac{5.74}{Re^{0.9}} + \frac{\epsilon/D_h}{3.7} \right) \right]. \tag{4.31}$$

For laminar flow ($Re < 2300$), a much simpler expression can be derived theoretically,

$$f = \frac{64}{Re}. \tag{4.32}$$

The pressure drops from fittings, valves, and duct area change are found using an equation
similar to Eq. (4.28):

$$\Delta P = K \frac{\rho U^2}{2} \tag{4.33}$$

where K is a fitting loss coefficient (listed in Table 4.1).

Table 4.1 Fitting loss coefficients

Type	K
Well rounded entrance	0.05
Contraction	0.38
Expansion	0.1
Short radius right angle bend	0.1
Diffuser	0.1 0.2
Tee (straight)	0.5
Tee (turn)	1.2

Source Adapted from *2017 ASHRAE Handbook—Fundamentals*, Chap. 3

The overall pressure drop in a piping or duct system is the sum of the friction and the fitting losses,

$$\Delta P = \left(f \frac{L}{D_h} + \sum K \right) \frac{\rho U^2}{2}. \tag{4.34}$$

From the continuity equation, $\dot{V} \approx \rho U$, so the overall system pressure drop scales with the square of the volumetric flow rate $\dot{V}$,

$$\Delta P \propto \dot{V}^2. \tag{4.35}$$

Example 4.3   *Pressure Drop in Piping Systems*

Cooling water flows from a central plant to an office building through a 12 cm diameter pipe which is $L = 100$ m in length. The pipe has two right-angle bends ($K = 0.1$) and a check valve ($K = 2.0$). The pipe roughness $\epsilon = 6.0 \times 10^4$ cm. The cooling water temperature is $T = 6\,°C$ and flow rate is 12 L/s (0.12 m³/s). Is the flow turbulent? What is the pressure drop and required pump head?

Solution

The mean velocity U in the pipe is

$$U = \frac{\dot{V}}{A}$$

$$= \frac{0.012}{\frac{\pi}{4}(0.12)^2} = 1.06\,\text{m/s}.$$

The dynamic viscosity μ is

$$D = (T - 8.435) = 6 - 8.435 = -2.435$$

$$\mu = \frac{0.1}{2.1482[D + (8078.4 + D^2)^{1/2}] - 120}$$

$$= 1.47 \times 10^3 \text{ Pa-s}.$$

The Reynolds number is

$$Re = \frac{\rho_w U D_h}{\mu}$$

$$= \frac{(1000)(1.06)(0.12)}{1.47 \times 10^3} = 86,390.$$

The flow is fully turbulent.
The friction factor for turbulent flow is

$$\frac{1}{f^{1/2}} = 0.8687 \left[\ln \left(\frac{5.74}{Re^{0.9}} + \frac{\epsilon/D_h}{3.7} \right) \right].$$

Upon substitution of the pipe parameters, $f = 0.0187$. The pressure drop in the piping system is thus

$$\Delta P = \left[f \frac{L}{D_h} + \sum K \right] \frac{\rho_w U^2}{2}$$

$$= \left[(0.0187) \frac{100}{0.12} + 2 + 2(0.1) \right] \frac{(1000)(1.06)^2}{2}$$

$$= [15.6 + 2.2] (561.8)$$

$$= 9.987 \times 10^3 \text{ Pa}.$$

The dominant loss mechanism is the pipe friction, which produces a pressure drop per unit length of about 100 Pa/m. The required pump head H_p is

$$H_p = \frac{\Delta P}{\rho g}$$

$$= \frac{9.987 \times 10^3}{(1000)(9.81)}$$

$$= 1.02 \text{ m}.$$

Note: For water distribution systems, per ASHRAE recommendations, the fluid velocity should be limited to no more than 1.2 m/s for pipes less than 5 cm in diameter, and for larger sizes, the pressure drop per unit length should be limited to no more than 400 Pa/m. Greater velocities will have noise and internal erosion issues.

4.7   Pump Work and Power

For an ideal pump, with no heat loss, the open system energy equation is

$$\dot{W}_{ideal} = \dot{m}\Delta h = \dot{m}\Delta(u + Pv).$$

(4.36)

We assume $v \simeq$ constant, since the fluid-specific volume, i.e., density, changes in air and water cooling systems are relatively small. Assuming that the internal energy u, i.e., temperature, of the working fluid is also constant, then

$$\Delta(u + Pv) \simeq v\Delta P = \frac{\Delta P}{\rho}$$

(4.37)

and

$$\dot{W}_{ideal} = \dot{m}\frac{\Delta P}{\rho} = \dot{V}\Delta P.$$

(4.38)

In water or air cooling systems, the ideal pumping power $\dot{W}_{ideal}$ is thus the product of the volumetric flow rate and the pressure rise across the pump.

The pump efficiency η_p is defined as the ideal work of the pump divided by the actual pump work,

$$\eta_p = \frac{W_{ideal}}{W_{actual}} = \frac{\dot{W}_{ideal}}{\dot{W}_{actual}}.$$

(4.39)

Example 4.4   Pump Power
If the pump efficiency in Example 4.3 is $\eta_p = 0.8$, what is the input power to the pump?

Solution
The ideal pump power is
$$\dot{W}_{ideal} = \dot{V}\Delta P$$
$$= (0.012)(9.987 \times 10^3)$$
$$= 119.8\,\text{W}.$$

The actual pump power is
$$\dot{W}_{actual} = \frac{\dot{W}_{ideal}}{\eta_p}$$
$$= \frac{119.8}{0.80}$$
$$= 149.7\,\text{W}.$$

4.8   Air Flow in HVAC Systems

In buildings, a Heating, Ventilating, and Air Conditioning (HVAC) system is used to supply conditioned air to rooms or zones, and to return exhaust air back to the outside environment. A schematic of a simple one-zone HVAC system is shown below in Fig. 4.2. The supply air to the zone is composed of both outside and return air. The heating/cooling coil cools and dehumidifies the supply air to specified temperature and humidity levels. The supply fan delivers this conditioned air to the occupied zone to meet the cooling load. The supply air flow rate and temperature are chosen to maintain acceptable thermal comfort and air quality in the occupied zone.

An exhaust fan is used to remove hot and humid air from the occupied zone. The exhaust air fan will cause outdoor air to infiltrate unless the outdoor air fan supplies a quantity of outdoor air equal to the exhausted and exfiltrated air. Buildings are usually pressurized to reduce or prevent infiltration. It is customary to supply about 5–20% more outdoor air than exhaust air to ensure pressurization. The plenum space between the room ceiling panels and the ceiling is frequently used as an exhaust return duct. To reduce the energy consumption of the system, some portion of the exhaust air is then recirculated. Mixing boxes are used to mix return air with the primary supply air at a location near the room diffuser. The purpose of mixing boxes is to maintain a constant volume of air in the occupied space.

Constant volume ventilation systems vary the supply air temperature and keep the flow rate of supply air constant. The amount of outdoor air brought into the building is constant at all times. The outdoor and return dampers are set at a minimum position, so the supply volume and percentage of outdoor air do not change. Outdoor air can be delivered separately to each occupied zone using a number of independent constant volume systems.

Variable air volume (VAV) systems meet changes in the space cooling/heating load by varying the supply air flow rate. In general, the VAV system maintains the supply air at a constant supply temperature and reduces or increases the air flow between preset maximum and minimum flows to meet the space loads.

The supply air flow rate can be found from an open system sensible energy balance on the occupied zone,

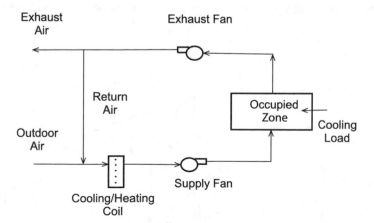

Fig. 4.2 Simple one-zone HVAC system schematic

$$\dot{V} = \frac{\dot{L}}{\rho c_p \Delta T}. \tag{4.40}$$

For example, if the cooling load $\dot{L} = 15$ kW, the occupied zone temperature $T_{zone} = 22\,^\circ$C, the supply temperature $T_{duct} = 10\,^\circ$C, the air density $\rho = 1.18$ kg/m^3, and air specific heat $c_p = 1.0$ kJ/kg -K, then

$$\dot{V} = \frac{\dot{L}}{\rho c_p \Delta T} = \frac{15}{(1.18)(1.0)(22 - 10)} = 1.06\,\mathrm{m}^3/\mathrm{s}.$$

If the latent loads are to be included, the energy balance is based on the enthalpies of moist air,

$$\dot{V} = \frac{\dot{L}}{\rho \Delta h}. \tag{4.41}$$

The pressure losses at a given flow rate due to friction and fittings are matched by the pressure increase provided by the supply fan. A simple duct-fan pressure profile is shown in the schematic of Fig. 4.3. Note that the outside air drawn in by the fan can be at sub-atmospheric pressure due to duct friction, the duct friction loss is linear, and the duct fitting loss occurs at a specific location. Balancing dampers are used to equalize zone pressures and produce the desired flow rates.

4.9   HVAC Duct Networks

With HVAC systems in multi-zone buildings, the duct diameter D is a critical design variable. As indicated by the duct schematic in Fig. 4.4, the duct layout in a multi-zone building is no longer a single duct, instead the duct system is a complex geometric network with numerous legs to the conditioned zones. The length L of each leg can be deduced from the architectural

Fig. 4.3 Duct-fan pressure profile

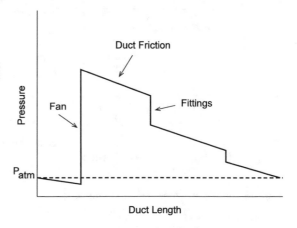

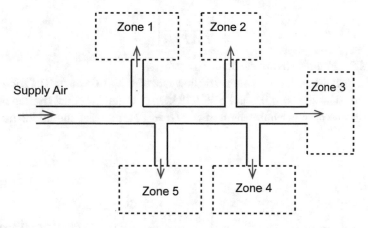

Fig. 4.4 Schematic of duct network

layout of the zones or rooms in the building. The volumetric flow rate of air $\dot{V}$ to each zone is found from the ventilation requirements to meet the load and supply the necessary outdoor air flow.

The selection of duct or pipe size is a trade-off between pumping power, noise, and cost. As we shall see, the pressure drop and thus pumping power decrease with the fifth power of the pipe diameter, and the pipe cost increases approximately linearly with diameter. There are also space limitations that also need to be considered.

The two most common duct network design methods for air distribution systems are *equal friction* and *static regain*. With the *equal friction* method, a friction pressure loss per unit length is chosen. Low-velocity systems have a pressure drop per unit length of about 1.0 Pa/m, and high-velocity systems have a pressure drop per unit length of about 5.0 Pa/m. The pressure drop equation, Eq. (4.34), and friction equation, Eq. (4.31), are solved iteratively for the duct diameter in each of the duct legs. The air velocity in each duct is known from the continuity equation, Eq. (4.15). The total fan power is given by the sum of the fan power required for each leg.

If the variation in friction factor and the fitting losses are relatively small, we can use the equal friction method to solve directly for the duct diameter D. Rearranging Eq. (4.34),

$$D = f \frac{\rho U^2}{2} \frac{1}{\Delta P / L}.$$ (4.42)

From the continuity equation, Eq. (4.15), we can write

$$U^2 = \frac{1}{D^4} \left(\frac{4}{\pi} \dot{V} \right)^2.$$ (4.43)

Upon substitution into Eq. (4.42),

$$D = \left[\frac{8}{\pi^2} \frac{\rho f}{\Delta P/L}\right]^{1/5} \dot{V}^{2/5}. \tag{4.44}$$

Example 4.5   Equal Friction Method

One leg of a duct network has a volumetric flow rate of $\dot{V} = 1.06\ \text{m}^3/\text{s}$. The friction factor in the duct is constant at $f = 0.02$, and $\rho = 1.18\ \text{kg/m}^3$. Using the equal friction method with a design pressure drop per unit length of $\Delta P/L = 1\ \text{Pa/m}$, what should the duct diameter D be?

Solution

From Eq. (4.44),

$$D = \left[\frac{8}{\pi^2} \frac{\rho f}{\Delta P/L}\right]^{1/5} \dot{V}^{2/5} = \left[\frac{8}{\pi^2} \frac{(1.18)(0.02)}{1}\right]^{1/5} (1.06)^{2/5} = 0.46\ \text{m}.$$

The duct velocity in this example is

$$U = \frac{1}{D^2}\left(\frac{4}{\pi}\dot{V}\right) = \frac{1}{(0.46)^2}\left(\frac{4}{\pi}1.06\right) = 6.3\ \text{m/s}.$$

Note: to reduce duct noise, air conditioning ducts are typically sized so that the duct air velocity is less than 10 m/s.

The *static regain* method is one that keeps the duct static pressure unchanged throughout the system. This is accomplished by successively reducing the velocity in the ducts. Ducts are sized so that the pressure drop in one duct section is balanced by the pressure rise in the upstream section.

4.10   Fan Performance

There are a number of relationships among fan (or pump) performance parameters called *fan laws* that are useful for predicting the change in the fan performance as the fan speed changes. These fan laws relate the volumetric flow rate $\dot{V}$, pressure rise ΔP, and power $\dot{W}$ to the fan speed. The fans used in air distribution systems are volume flow devices, as the flow is produced by rotating blades that sweep a given volume of air in every revolution, producing a volumetric flow rate proportional to the rotational speed. Therefore, for a change in fan speed from N_1 to N_2, the volumetric flow rate change scales linearly with the fan speed

$$\frac{\dot{V}_2}{\dot{V}_1} = \frac{N_2}{N_1}. \tag{4.45}$$

The pressure rise is proportional to the square of the fluid velocity (Eq. (4.35)), which scales with the fan speed, so the pressure rise scales with the square of the fan speed:

$$\frac{\Delta P_2}{\Delta P_1} = \left(\frac{N_2}{N_1}\right)^2. \tag{4.46}$$

The fan power is the product of the volume flow rate and pressure rise (Eq. (4.38)), so the fan power scales with the cube of the fan speed:

$$\frac{\dot{W}_2}{\dot{W}_1} = \left(\frac{N_2}{N_1}\right)^3. \tag{4.47}$$

Example 4.5 Fan Speed

An air conditioning system fan is delivering 4.0 m^3/s at a speed of 900 rpm and requires 6.0 kW of power. With an increase in the cooling load, there is a need to increase the air flow rate to 5.0 m^3/s. At what speed should the fan be operated, and what is the required power?

Solution

Using the fan laws presented above,

$$N_2 = N_1 \left(\frac{\dot{V}_2}{\dot{V}_1}\right) = 900 \left(\frac{5.0}{4.0}\right) = 1125 \, \text{rpm}$$

$$\dot{W}_2 = \dot{W}_1 \left(\frac{N_2}{N_1}\right)^3 = 6 \left(\frac{1125}{900}\right)^3 = 11.7 \, \text{kW}.$$

Note that for a 25% increase in flow rate, the power requirement is almost doubled.

Representative fan performance curves are shown in Fig. 4.5, in which the fan delivery pressure is plotted versus the flow rate. For a fixed fan speed, as the flow rate is increased, the pressure rise decreases. As the fan speed is increased, the fan curve moves upward as both the pressure rise and the flow rate increase.

The operating point of an air distribution system is found by the intersection of the system curve and the fan performance curve. As shown in Fig. 4.6, at the operating point, the system flow rate is at the location on the fan curve where the pressure rise of the fan matches the pressure drop of the duct system.

Example 4.6 Fan Power

What is the operating point and minimum power consumption for the fan-duct system characterized by the following equations? (ΔP in Pa, $\dot{V}$ in m^3/s)

$$\text{Fan}: \quad \Delta P = 85 - 7(\dot{V}^2 - \dot{V})$$

$$\text{Duct}: \quad \Delta P = 10\dot{V}^2.$$

Fig. 4.5 Fan performance
curves

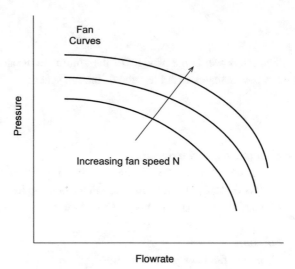

Fig. 4.6 System operating
point

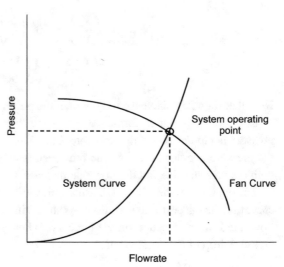

Solution

Equating the fan pressure rise with the duct pressure drop,

$$\Delta P_{fan} = \Delta P_{duct}$$
$$85 - 7(\dot{V}^2 - \dot{V}) = 10\dot{V}^2$$
$$17\dot{V}^2 + 7\dot{V} - 85 = 0.$$

The positive root of the quadratic is $\dot{V} = 2.04$ m^3/s, resulting in a pressure drop of 70.1 Pa.
The minimum power consumption is

$$\dot{W}_{ideal} = \dot{V} \Delta P$$
$$= (2.04)(70.1)$$
$$= 143 \, \text{W}.$$

4.11   Cooling Coil Fluid Flow

A cooling coil is a tubular multipass crossflow heat exchanger that both cools and dehumidifies an air stream. Cooling coils have a large face area on the air side of the heat exchanger, and a smaller flow area on the liquid side. In large HVAC systems with a central chiller, chilled water is circulated to numerous cooling coils located in different building zones. The chilled water flows inside the tubes and the supply air flows over the tubes, with heat transfer from the hot supply air to the cooling water.

If the temperature of the air side of the coil surface is above the air dew point temperature, there will not be any condensation of the water vapor in the air, and the tube surface will remain dry. Since there is only sensible cooling of the air in this case, the cooling coil performance can be determined using a *dry* analysis. However, if the temperature of the air side of the coil surface is below the air dew point temperature, there will be condensation of the water vapor in the air onto the tube, and the tube surface will be wet. The performance of a *wet* coil is found from convection heat and mass transfer analyses that include both sensible and latent cooling. The additional parameters include the enthalpy of the saturated air–water vapor mixture at the tube surface, and the enthalpy of the air stream at a distance away from the tube surface.

The mean temperature and humidity change for moist air flowing through a wet cooling coil are shown in Fig. 4.7. The moist air that initially contacts the cold coil will reach and then drop below the dew point while the mean air temperature remains above the dew point. Therefore, the absolute humidity of the air will continually drop as the air cools and eventually reaches the saturation line.

For determination of building air conditioning needs, the conditioned air is assumed to leave the cooling coil at 12.8 °C (55 °F) with a relative humidity of about 0.90, and the liquid entering the tube side of the cooling coil is typically assumed to be 6.7 °C (445 °F). Since the dew point for air at 23.9 °C (755 °F) and 50% relative humidity is 12.8 °C (555 °F), there will be some dehumidification of the occupied space with a cooling coil.

With a direct expansion (DX) cooling coil, the liquid refrigerant flows directly in the tubes of the cooling coil, and evaporates into the vapor phase, cooling the air stream. A finned tube heat exchanger is used in DX systems. Heat transfer from a hot air stream originating from the cooling load will evaporate the liquid refrigerant passing through the tube. Air-cooled evaporators operating below 0 °C will accumulate frost from water vapor solidifying on the coils, so arrangements must be made for periodic frost removal, such as electric heaters or reverse cycling.

Fig. 4.7 Temperature and
humidity changes for moist air
flow through a cooling coil

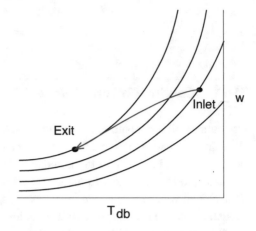

4.12   Room Air Flow

There are two main types of room air flow supply schemes—mixed and displacement. With
mixed ventilation, a high-velocity jet of conditioned air from the ceiling or wall is used
to mix the air in the room, producing a uniform temperature throughout the space. Mixed
ventilation is the dominant room air flow method in the United States. With displacement
ventilation, low-velocity jets of supply air from the floor create thermal stratification and
displace the room air to the ceiling out of the occupied zone. Displacement ventilation has
very high effectiveness for contaminant removal, since the mixing of the supply air with the
room air is minimized (Sandberg 1989). For both methods, a return grill is used to keep the
mass of air in the room constant.

With both supply schemes, the introduction of the conditioned duct air to the occupied
zone is performed by a device called a diffuser. The diffuser is located on a ceiling, wall, or
floor, and its function is to convert some portion of the duct static pressure to an increase
in the flow velocity and thus increase the momentum of the supply flow. The outlet diffuser
temperatures for cooling applications range from 10 to 13 °C. With the advent of ice storage
systems in commercial buildings, the outlet diffuser temperatures can be lowered to 6 °C,
which reduces the required HVAC duct size and volumetric air flow rate (Kirkpatrick 1996).

Slot and Radial Diffusers for Mixed Ventilation
Diffusers are grouped into two major categories, linear and radial. Two common types of
linear diffusers are slot and jet diffusers. Common types of radial diffusers include circular,
square, crossflow, swirl, and perforated. The most prevalent diffusers in HVAC applications
are slot and circular. The slot diffuser is a ceiling-mounted diffuser that produces a linear
jet from a line source. The exit air flow can be configured to produce a one-way or two-way
flow. Due to curved blades at the exit, the air flow from a slot diffuser deflects toward and
adheres to the ceiling, forming a wall jet. Slot diffusers are typically 0.6–1.5 m long, with

slot widths varying from 12 to 24 mm. The sideways or lateral spread of the jet is very small compared to the downward spread. A radial diffuser is a ceiling-mounted diffuser that produces a circular jet. The diffuser is composed of concentric air passages that can be round, square, or rectangular in shape, with an internal geometry that produces a radial or crossflow pattern.

The mixing of the diffuser air jet with the room air will cause a decrease in the jet centerline velocity, while producing nearly uniform thermal conditions in the occupied space. The distance that the jet travels during this mixing process is called the *throw* of the jet. Mathematically, the throw is defined as the distance from the diffuser outlet to the point where the jet velocity has decreased to a terminal value, typically, 0.25 or 0.50 m/s. The throw is a significant design variable as it is an indicator of proper room air mixing (Koestel 1957).

There are two limits to acceptable throw values. If the throw is too short, relative to the room length, the room air will be stagnant and stratified, due to poor mixing. If the throw is too long, the occupied zone will be too drafty. In addition, diffusers for both heating and cooling applications should have a sufficient throw to properly mix with the room air during both the heating and cooling seasons.

Air Diffuser Performance Index

The thermal comfort performance of diffusers is quantified by the Air Diffuser Performance Index (ADPI) (Miller 1979). The ADPI is defined as the percentage of measurement locations in the occupied zone of a space that have effective draft temperatures between $-1.7\,^{\circ}$C and $+1.1\,^{\circ}$C, and air speeds less than 0.35 m/s. These are conditions that most people working in an office will find comfortable. A draft is the undesired cooling of a room's occupants by air movement. The effective draft temperature is defined as the difference between the local air temperature and a room air temperature that would produce the same cooling effect as air flowing by a person. The equation for the effective draft temperature T_{ed} is

$$T_{ed} = (T - \overline{T}) - a(V - b) \tag{4.48}$$

where
$$T = \text{local air temperature } (^{\circ}\text{C})$$
$$\overline{T} = \text{occupied zone average temperature}(^{\circ}\text{C})$$
$$a = 8.010$$
$$b = 0.15\,\text{m/s}$$
$$V = \text{local air speed (m/s)}.$$

For example, if the local air temperature is $24\,^{\circ}$C and local air speed is 0.25 m/s, with an average occupied zone temperature of $22.2\,^{\circ}$C, then the local effective draft temperature would be $T_{ed} = -1.9\,^{\circ}$C. This is a condition that would be perceived as too cold.

Tests of diffusers over a wide range of loads, flow rates, and supply-room temperature differences have resulted in recommendations for the ratio of the throw to a room characteristic length, T/L, that result in a maximum value of ADPI, and a range of values of T/L where the ADPI is above a minimum 80% value. The room characteristic length L is shown in Fig. 4.8 for radial and slot diffusers.

Representative T/L values for slot and radial diffusers are given in Table 4.2. Note that the T/L range extends over a factor of six for slot diffusers and a factor of two for radial diffusers. These types of diffusers perform well over a wide range of flow rates, which is necessary when used with variable air volume (VAV) air conditioning systems.

Diffuser Selection

The type of diffuser selected depends on the preference of the architect and the building owner. Various types of diffusers are listed in manufacturer's catalogs and data sheets. As indicated in Table 4.3, the catalog data lists flow rate, throw, pressure drop, and noise criteria for a given diffuser model. For a known room geometry and cooling load, a diffuser should

Fig. 4.8 Room characteristic lengths (L)

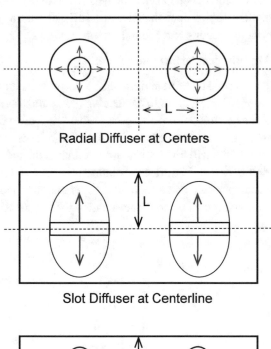

Radial Diffuser at Centers

Slot Diffuser at Centerline

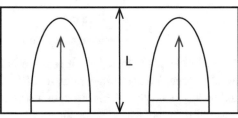

Slot Diffuser at Wall

Table 4.2 T/L ranges for slot and radial diffusers

Type	Characteristic length	T/L range	ADPI range
Slot	Distance to wall or symmetry plane	0.5–3.3	85–92
Radial	Distance to wall or symmetry plane	0.7–1.5	76–93

Source Adapted from *2017 ASHRAE Handbook—Fundamentals, Chap. 14*

be selected that will give an acceptable value of ADPI plus meet other criteria such as noise. With linear slot diffusers, a *rule of thumb* is that the sum of the active lengths of the diffusers be 30–70% of the length of the ceiling or wall.

The Noise Criteria (NC) parameter is a single number that represents the noise level of the diffuser. It is based on measurements of the sound level at various octaves. NC15 is defined as total silence. The recommended NC level for conference rooms and classrooms is NC 25–30 or below.

Example 4.7   Conference Room Diffuser Selection

A conference room 4.6×9.0 m has a cooling load $\dot{L} = 4.5$ kW and a design outdoor air ventilation rate of 0.21 m³/s. The supply air temperature is $10\,^{\circ}$C and the room set point temperature is $23\,^{\circ}$C. What are the slot diffuser design parameters?

Solution

The maximum supply air flow rate needed to meet the cooling load is

$$\dot{V} = \frac{\dot{L}}{\rho c_p \Delta T} = \frac{4.5}{(1.18)(1.0)(23 - 10)} = 0.29 \, \text{m}^3/\text{s}.$$

As specified, the minimum flow rate is 0.21 m³/s with 100% outside air. A proposed slot diffuser is listed in Table 4.3. The diffuser is 122 cm long, with a 2.5 cm slot width. It is to be located along one side of the longer wall, so the characteristic length L is 4.6 m. From

Table 4.3 Representative slot diffuser catalog data

Flow rate (m³/s)	Throw (m)	Static pressure (Pa)	Noise criteria (NC)
0.07	7.0	20.1	26
0.08	7.6	27.6	28
0.09	7.9	36.6	30
0.10	8.5	46.3	32

Note Table data is for a one-way slot diffuser 120 cm long, with a 2.5 cm slot width

Table 4.2, the range of throw/characteristic length T/L is 0.5–3.3, so the range of acceptable throw values is 2.3–15.2 m.

Following the typical *rule of thumb*, three diffusers will be used, for an active length of 3.7 m, about 40% of the longer wall length. The maximum flow through each diffuser is therefore $0.29/3 = 0.098$ m³/s, and the minimum flow is $0.21/3 = 0.070$ m³/s. From the data in Table 4.3, the slot diffuser has a throw of about 7.9 m at maximum flow, and a throw of about 7.3 m at the minimum flow, which meet the throw requirements over the flow rate range. The static pressure drop is about 35 Pa at the maximum flow rate, producing acceptable noise criteria of about 30.

4.13   Mixing Boxes

Mixing boxes are used to mix return plenum air with the primary supply air at a location near the room diffuser. The purpose of mixing boxes is to maintain a constant volume of air in the occupied space. As the cooling load deceases in a space, the mixing boxes will increase the amount of plenum air mixed with the primary supply air. The temperature of the air supplied to the space will rise, since the cooling load has decreased, but the volume flow rate to the occupied space will be kept as constant as possible. This will ensure adequate mixing of the room air by the supply air to maintain acceptable thermal comfort conditions.

As shown in Fig. 4.9, terminal mixing boxes are classified as induction, parallel, and series. The induction boxes operate by entrainment of plenum air. A primary modulating damper in the induction box lowers the primary air flow static pressure, which induces higher pressure plenum air into the induction box. The induced air is supplied directly from the return air path through an induction damper. The induction rate is non-linear since the local static pressure is proportional to the square of the air flow rate. When full cooling of a room is required, the ventilation system maintains the primary air volume at the induction box and closes the induction damper.

Series boxes have a fan operating in series with the main air handler. Plenum air is drawn by the series fan into the mixing box and mixed with the colder primary supply air upstream of the fan. The fan is sized for the air quantity that would be required for 12 °C (555 °F) supply air. The series fan runs continuously, supplying the diffusers with a nearly constant flow rate of air. Control is achieved with an inlet air damper.

Parallel boxes have a fan operating in parallel with the main air handler. The fan is located outside the primary air stream, and draws plenum air into the mixing box. Parallel boxes are useful in exterior or perimeter spaces that require heating, while interior zones require cooling. After the primary air has been reduced to the minimum required for ventilation, and the zone temperature has decreased to the heating set point, the fan is used to provide heat using plenum air. Parallel mixing boxes require a higher static inlet pressure, typically 124 Pa (0.5 in w.g.) than series boxes since the primary air is not drawn by a fan in the mixing box.

Fig. 4.9 Induction, series, and parallel mixing boxes

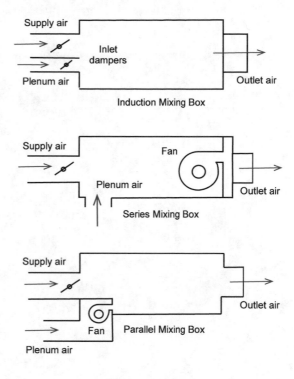

4.14 Further Reading

Additional information about fluid flow in pipes and ducts is given in the textbook by White (2015).

References

Kirkpatrick A, Elleson J (1996) Design guide for cold air distribution. ASHRAE, Atlanta, Georgia

Koestal A (1957) Jet velocities from radial flow outlets. ASHAE Transactions 63:505

Miller P (1979) Design of room air diffusion systems using the air diffusion performance index (ADPI). ASHRAE J 10:85

Perry R, Green D, and Malrnay J (1999) Perry's Chemical Engineers' Handbook. McGraw Hill, New York

Sandberg M, Blomqvist C (1989) Displacement ventilation in office rooms. ASHRAE Transactions 95(2):1041-1049

White F (2015) Fluid Mechanics. McGraw Hill, New York

Heat Transfer in HVAC Systems

5

5.1 Introduction

In this chapter we examine the heat transfer that occurs in refrigeration and air conditioning systems. Heat transfer is a thermal energy transfer driven by temperature differences.
Thermal energy transfer has been grouped into three mechanisms: conduction, convection,
and radiation. Conduction heat transfer results from molecular-level kinetic energy transfer
in solids, liquids, and gases. Convection heat transfer is thermal energy transfer between
a solid and a fluid. The physical mechanism of radiation heat transfer is energy transfer
by electromagnetic waves or photons emitted from one surface and absorbed by another
surface. Most heat transfer situations involve more than one mode of heat transfer, such as
combined conduction and convection. We review heat transfer resistance networks, which
are useful approaches for modeling multi-mode situations.

A major component of the cooling load of a building is due to the heat transfer into the
building, which in turn depends on the building temperature, the environment temperature,
and the physical properties of the building walls, windows and roof. In addition, heat transfer
to the cool supply air flowing through air conditioning ductwork results in a temperature
increase of the air flow. This duct heat gain is added to the space cooling load, increasing
the required supply air volume.

5.2 Conduction Heat Transfer

Conduction heat transfer results from molecular-level kinetic energy transfer in solids, liquids, and gases. In accordance with the second law of thermodynamics, the thermal energy
flow through the material is in the direction of decreasing temperature. As shown in Fourier's
equation, Eq. (5.1), the heat flow per unit area $\dot{Q}/A$ is directly proportional to the temperature

A. T. Kirkpatrick, *Introduction to Refrigeration and Air Conditioning Systems*,
Synthesis Lectures on Mechanical Engineering,
https://doi.org/10.1007/978-3-031-16776-8_5

Table 5.1 Thermal conductivity of various materials

Material	k (W/m K)
Air	0.027
Glass fiber insulation	0.036
Wood	0.10
Glass	0.78
Concrete	1.73
Steel	45
Aluminum (2024-T6)	151

Source Adapted from *2017 ASHRAE Handbook—Fundamentals*, Chap. 4

gradient with a constant of proportionality defined as the thermal conductivity k.

$$\frac{\dot{Q}}{A} = -k\frac{dT}{dx} \tag{5.1}$$

For steady conduction through a solid, such as a wall in a building, with constant k and thickness L, integration of Eq. (5.1) across the wall yields

$$\frac{\dot{Q}}{A} = k\frac{T_1 - T_2}{L} \tag{5.2}$$

The thermal conductivity k of various materials used in buildings and cooling systems is presented in Table 5.1. Note the wide range of thermal conductivity values, which vary by several orders of magnitude. Gases have low densities with corresponding low thermal conductivities. Insulation materials have voids filled with air and also have low thermal conductivities. Metals have much greater thermal conductivities due to their electron mobility. As discussed in Chapter Four, the thermal conductivity of most non-metals increases with temperature. With the relatively small temperature differences characteristic of HVAC systems, the thermal conductivity of the components of these systems is typically assumed to be constant.

5.3   Convection Heat Transfer

Convection heat transfer is thermal energy transfer between a solid and a fluid. It is associated with large scale motion of a fluid past a relatively warmer or cooler surface. Convection processes are grouped into different classes, each with their own correlation equations. With forced convection, the fluid motion results from an external pressure difference, such as a fan or pump. With natural or free convection, the fluid motion is caused by density differences

Table 5.2 Range of heat transfer coefficients

Convection Mode	$\mathbf{h}$ (W/m^2 K)
Natural convection (air)	5–20
Forced convection (air)	20–300
Forced convection (water)	200–5000
Boiling (water)	5000–20000
Condensation	5000–10000

in the fluid created by the warmer or cooler adjacent surface. For example, a vertical hot surface will heat the air around it, producing natural convection flow past the surface.

As shown by Newton's convection equation, Eq. (5.3), the thermal energy flow rate per unit area is linearly proportional to the fluid-surface temperature difference. The constant of proportionality, $\mathbf{h}$, is defined as the heat transfer coefficient, with units W/m^2 K. Typical order of magnitude values of the heat transfer coefficient are given in Table 5.2. Note that coefficients span a wide range of magnitudes. Depending on the problem being addressed, the heat transfer coefficient can be used to determine localized thermal conditions, and it can also be integrated over the surface to provide the overall heat transfer.

$$\frac{\dot{Q}}{A} = \mathbf{h}(T_1 - T_2) \tag{5.3}$$

The heat transfer coefficient $\mathbf{h}$ depends on the physical properties of the fluid, the geometry of the solid surface, and the fluid velocity. These parameters are formed into dimensionless groups, namely, the Nusselt, Reynolds, Rayleigh, and Prandtl numbers, defined as follows:

$$Nu = \text{Nusselt number} = \frac{\mathbf{h}L}{k}$$

$$Re = \text{Reynolds number} = \frac{UL}{\nu}$$

$$Ra = \text{Rayleigh number} = \frac{g\beta\Delta T L^3}{\nu\alpha}$$

$$Pr = \text{Prandtl number} = \frac{\nu}{\alpha}$$

The parameter U is the fluid velocity, and L is a length scale. The physical properties in the above terms are

$$k = \text{thermal conductivity (W/m K)}$$
$$v = \text{kinematic viscosity (m}^2/\text{s)}$$
$$\alpha = \text{thermal diffusivity (m}^2/\text{s)}$$
$$\beta = \text{coeff. of thermal expansion}$$
$$g = \text{gravity (9.81 m/s}^2)$$

The Nusselt number represents a non-dimensional heat transfer coefficient. The Nusselt number is named after Wilhelm Nusselt (1882–1957), a German engineering professor who made many contributions to heat transfer, primarily in dimensional analysis, condensation, and heat exchangers. The Reynolds number is a ratio of the fluid inertia to viscosity, the Rayleigh number is a ratio of the fluid buoyancy to viscosity, and the Prandtl number is a ratio of the the ratio of the diffusion of momentum to the diffusion of thermal energy. The Prandtl number is named after Ludwig Prandtl (1875–1953), a German engineering professor who made many significant contributions to fluid mechanics and aerodynamics.

The fluid motion in convection can be characterized as laminar or turbulent, depending on the value of the Reynolds number (forced convection) or Rayleigh number (free convection) of the flow.

Experimental convection data is correlated using the dimensionless relationships given below in Eq. (5.4).

$$Nu = a \, Re^b \, Pr^c \quad \text{forced convection}$$
$$Nu = a \, Ra^b \, Pr^c \quad \text{free convection} \tag{5.4}$$

For example, the correlation for turbulent flow in a pipe with $Re > 10,000$, with the fluid being heated is

$$Nu = 0.023 \, Re^{4/5} \, Pr^{0.4} \tag{5.5}$$

It is helpful to see the convection equations in dimensional form as well as non-dimensional form, by substituting numerical values for physical properties at $20\,°\text{C}$. For external forced convection flow over plane surfaces, such as air flowing over a building wall or roof with a characteristic length L, if $UL < 1.4 \text{ m}^2/\text{s}$, i.e., the Reynolds number $<$ 500,000, the laminar air flow correlation is

$$h = 2.0 \left(\frac{U}{L} \right)^{1/2} \tag{5.6}$$

and if $UL > 1.4 \text{ m}^2/\text{s}$, the turbulent air flow correlation is

$$h = 6.2 \left(\frac{U^4}{L} \right)^{1/5} \tag{5.7}$$

For natural convection air flow over horizontal pipes of diameter D, such as flow over a heat exchanger coil, if $D^3 \Delta T < 1.0 \text{ m}^3 \,°\text{C}$, the laminar flow correlation is

$$h = 1.32 \left(\frac{\Delta T}{D}\right)^{1/4} \tag{5.8}$$

and if $D^3 \Delta T > 1.0 \text{ m}^3 \, °C$, the turbulent flow correlation is

$$h = 1.24 \Delta T^{1/3} \tag{5.9}$$

For the case of internal forced convection turbulent flow, the heat transfer coefficient for fully developed air flow in a duct is

$$h = 8.8 \left(\frac{U^2}{D_h}\right)^{1/5} \tag{5.10}$$

The parameter D_h is the hydraulic diameter. It is defined as

$$D_h = \frac{4A}{P} \tag{5.11}$$

where A is the cross sectional area, and P is the duct or pipe perimeter. A round duct has a hydraulic diameter D_h equal to the actual diameter D.

The heat transfer coefficient for fully developed turbulent flow of water in a pipe has a temperature as well as a velocity dependence,

$$h = 3580(1 + 0.015T)\left(\frac{U^4}{D_h}\right)^{1/5} \tag{5.12}$$

In Eqs. (5.10) and (5.12) above, the velocity is in m/s, the hydraulic diameter is in cm, and the temperature is in °C. Additional forced and natural convections correlations for other geometries are given in standard heat transfer textbooks.

5.4 Thermal Resistance

Most heat transfer situations involve more than one mode of heat transfer, such as combined conduction and convection. A very useful approach for modeling such multi-mode situations is the use of a heat transfer resistance network. With a resistance network, composite serial and parallel resistances can be combined into a single resistance, simplifying the heat transfer analysis. The overall heat flow is modeled as one-dimensional from the hotter to the cooler surface.

We define thermal resistance R as the ratio of a temperature difference ΔT to the heat flow $\dot{Q}$ across the temperature difference, with units °C/W or K/W.

$$R = \frac{\Delta T}{\dot{Q}} \tag{5.13}$$

For conduction heat transfer, since

$$\dot{Q} = -kA\frac{\Delta T}{\Delta x} \tag{5.14}$$

The conduction resistance value R_{cond} is

$$R_{cond} = \frac{\Delta x}{kA} \tag{5.15}$$

Similarly, for convection heat transfer, since

$$\dot{Q} = \mathbf{h}A\Delta T \tag{5.16}$$

The convection resistance value R_{conv} is

$$R_{conv} = \frac{1}{\mathbf{h}A} \tag{5.17}$$

A *commercial* r-value is defined by multiplying the resistance R by the area A

$$r = RA \tag{5.18}$$

Building codes specify minimum r-values for walls, roofs, and floors. Depending on the climate, the wall r-values range from 3.5 to 5.0, and the roof r-values range from 5.0 to 10. The conversion factor from English units to metric units is 0.176. For example, a r-20 wall in English units (ft^2 °F hr/Btu) is r-3.5 in metric units (m^2 °C/W).

For a plane wall with convection on either side, and conduction through the wall, a series resistance network, with nodes for the air temperatures and wall surface temperatures, is shown in Fig. 5.1. At steady state, the heat transfer rate per unit area from T_i to T_o is

$$\frac{\dot{Q}}{A} = \frac{T_i - T_o}{\sum r_i} \tag{5.19}$$

The sum of the three resistors is

$$\sum r_i = \frac{1}{\mathbf{h}_1} + \frac{\Delta x}{k} + \frac{1}{\mathbf{h}_2} \tag{5.20}$$

An overall heat transfer coefficient $\mathbf{U}$ is defined in terms of the total thermal resistance to heat transfer between two locations,

$$\mathbf{U} = \frac{1}{\sum r_i} = \frac{1}{\frac{1}{\mathbf{h}_1} + \frac{\Delta x}{k} + \frac{1}{\mathbf{h}_2}} \tag{5.21}$$

Accordingly,

$$\frac{\dot{Q}}{A} = \mathbf{U}(T_i - T_o) \tag{5.22}$$

where $T_i - T_o$ is the overall temperature difference. If there are area changes along the heat transfer path, then each resistance element should be divided by the local area.

Example 5.1 Building Wall Heat Transfer

As shown in Fig. 5.1, a building wall separates a room at a temperature $T_i = 25\,°C$ from the outside environment which is at a temperature $T_o = -10\,°C$. The heat transfer coefficient $\mathbf{h_1}$ on the interior side of the wall is 5 W/m² K, and on the exterior side $\mathbf{h_2}$ is 20 W/m² K. The wall is 15 cm thick, with an r-value of 3.5 m² °C/W. What is the heat transfer rate per unit area (W/m²) and the interior surface temperature T_1 of the wall?

Solution

The total resistance is

$$\sum r_i = \frac{1}{\mathbf{h_1}} + r + \frac{1}{\mathbf{h_2}} = \frac{1}{5} + 3.5 + \frac{1}{20} = 3.75$$

The overall heat transfer coefficient $\mathbf{U}$ is

$$\mathbf{U} = \frac{1}{\sum r_i} = \frac{1}{3.75} = 0.267 \text{ W/m}^2\text{K}$$

The heat transfer rate per unit area is

$$\frac{\dot{Q}}{A} = \mathbf{U}(T_i - T_o) = (0.267)[25 - (-10)] = 9.33 \text{ W/m}^2$$

The heat flow through the interior convective resistor is

Fig. 5.1 Series thermal resistance network

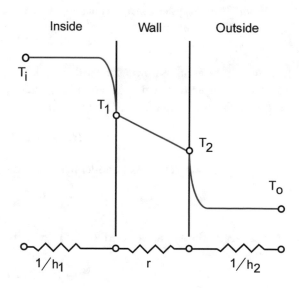

$$\dot{Q} = \mathbf{h}_1 A(T_i - T_1)$$

Solving for T_1,

$$T_1 = T_i - \frac{\dot{Q}}{\mathbf{h}_1 A} = 25 - \frac{9.33}{5} = 23.1\,\text{C}$$

5.5   Heat Transfer in Ductwork

Heat transfer to the cool supply air flowing through air conditioning ductwork results in a temperature increase of the air flow. This duct heat gain is added to the space cooling load, increasing the required supply air volume. The duct heat gain originates from the warmer surrounding air, supply and return fan inefficiencies, and mixing box fans.

5.5.1   Duct Heat Transfer

Duct heat gain from the surrounding warmer air depends primarily on the local temperature difference $(T_\infty - T)$ between the duct air and the air surrounding the duct, the duct insulation resistance, and the external and internal heat transfer coefficients. The latter terms can be grouped into an overall heat transfer coefficient U. Insulation on ductwork and other air distribution equipment serves to reduce heat gain to the supply air and to prevent condensation of ambient moisture. In general, with conventional supply air temperatures between 10 and 15 °C, a 25 mm glass fiber mat is used. For colder supply air temperatures between between 5 and 10 °C, thicker insulation (50 mm or greater) is recommended.

From the open system energy equation, the overall duct heat gain $\dot{Q}$ is

$$\dot{Q} = \dot{m}c_p(T_o - T_i) \tag{5.23}$$

where $\dot{m}$ is the air mass flowrate in the ductwork, T_o is the duct outlet temperature, T_i is the duct inlet temperature, and c_p is the air specific heat. For a duct section of length dx (see Fig. 5.2), the differential form of Eq. (5.23) is

$$d\dot{Q} = \dot{m}c_p dT \tag{5.24}$$

and the rate of convection heat transfer in the section dx is

$$d\dot{Q} = \frac{\dot{Q}}{A} P dx \tag{5.25}$$

The convective heat transfer from the warm environment to the cold air in the duct at a given location is

$$\frac{\dot{Q}}{A} = \mathbf{U}(T_\infty - T) \tag{5.26}$$

Fig. 5.2 Duct heat gain

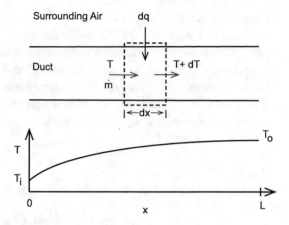

Upon substitution of Eq. (5.26)

$$d\dot{Q} = \dot{m}c_p dT = \mathbf{U}(T_\infty - T)P dx \tag{5.27}$$

Solving for the temperature gradient dT/dx,

$$\frac{dT}{dx} = \frac{P\mathbf{U}}{\dot{m}c_p}(T_\infty - T) \tag{5.28}$$

Defining the temperature difference θ between the environment and the local temperature,

$$\theta = T_\infty - T$$
$$d\theta = -dT \tag{5.29}$$

Separating variables and integrating along the length L of the duct from inlet (i) to outlet (o), assuming $\mathbf{U}$ is constant,

$$\int_{\theta_i}^{\theta_o} \frac{d\theta}{\theta} = -\frac{P\mathbf{U}}{\dot{m}c_p} \int_0^L dx$$

$$\ln\frac{\theta_o}{\theta_i} = -\frac{P\mathbf{U}}{\dot{m}c_p}L \tag{5.30}$$

$$\frac{\theta_o}{\theta_i} = \frac{T_\infty - T_o}{T_\infty - T_i} = \exp\left(-\frac{P\mathbf{U}L}{\dot{m}c_p}\right)$$

Solving for the outlet temperature T_o,

$$T_o = T_\infty + (T_i - T_\infty)\exp\left(-\frac{P\mathbf{U}L}{\dot{m}c_p}\right) \tag{5.31}$$

The temperature rise along the duct is not linear, and has an exponential-type profile.

Example 5.2 A/C Duct Heat Gain

An air conditioning duct has a volumetric flowrate $\dot{V} = 0.5 \text{ m}^3/\text{s}$. The inlet air temperature is $10.0\,°\text{C}$, and the surrounding air temperature is $25.0\,°\text{C}$. The duct is 400 mm wide by 200 mm high, 20 m long, and insulated with 25 mm thick glass fiber duct wrap. The duct insulation r-value is $0.53 \text{ Km}^2/\text{W}$, and exterior natural convection heat transfer coefficient is $9.2 \text{ W/m}^2\text{K}$. (a.) What is the forced convection heat transfer coefficient on the inside of the duct and the overall **U** value? (b.) What is the outlet air temperature and the rate of heat transfer to the duct air?

Solution

(a) The hydraulic diameter D_h of the duct is

$$D_h = \frac{4A}{P} = \frac{4(0.4)(0.2)}{2(0.4 + 0.2)} = 0.267 \text{ m}$$

The mean flow velocity U is

$$U = \frac{\dot{V}}{A} = \frac{0.5}{(0.4)(0.2)} = 6.25 \text{ m/s}$$

From Eq. (5.10), the interior heat transfer coefficient h_i is

$$h = 8.8 \left(\frac{U^2}{D_h}\right)^{1/5} = 8.8 \left(\frac{6.25^2}{0.267}\right)^{0.20} = 23.8 \text{ W/m}^2\text{K}$$

The overall **U** value is

$$\mathbf{U} = \frac{1}{\frac{1}{h_i} + r_{insul} + \frac{1}{h_o}}$$

$$= \frac{1}{\frac{1}{23.8} + 0.53 + \frac{1}{9.2}}$$

$$= 1.47 \text{ W/m}^2\text{K}$$

Note that the largest thermal resistance is due to the insulation layer, with much smaller resistances from the exterior natural convection, and the interior forced convection.

   (b) The duct perimeter is

$$P = 2(\text{height} + \text{width}) = 2(0.2 + 0.4) = 1.2 \text{ m}$$

The exponential parameter is

$$\frac{PUL}{\dot{m}c_p} = \frac{PUL}{\rho \dot{V} c_p} = \frac{(1.2)(1.47)(20)}{(1.18)(0.5)(1005)} = 5.95 \times 10^{-2}$$

The outlet temperature T_o is

$$T_o = T_\infty + (T_i - T_\infty)\exp\left(-\frac{PUL}{\dot{m}c_p}\right)$$

$$= 25 + (10 - 25)\exp\left(-5.95 \times 10^{-2}\right)$$

$$= 10.87°\,C$$

The heat gain to the duct air is

$$\dot{Q} = \rho \dot{V} c_p (T_o - T_i)$$

$$= (1.18)(0.5)(1005)(10.87 - 10.0) = 513\,W$$

5.5.2   Fan Heat Gain

Fan heat is a significant cooling load that must be included in the cooling load calculation, since the fans are typically located in the airstream. Fan efficiency η_{fan} for air-handling unit supply and return fans is generally between 50 and 70%, and motor efficiency η_{mot} generally between 80 and 90%. Fan-powered mixing boxes temper the primary supply air with recirculated room air before introduction to the space. The efficiency of the fractional horsepower motors used in mixing boxes is of the order of 35%.

Since the kinetic energy of the airflow eventually ends up as viscous dissipation in a zone downstream for motors and fans located in the airflow, it is customary to equate the fan heat gain with the power supplied to the motor and fan.

$$\dot{Q}_{fan} = \dot{W}_{act}$$

$$= \frac{\dot{W}_{ideal}}{(\eta_{fan})(\eta_{mot})} \tag{5.32}$$

$$= \frac{\dot{V}\Delta P}{(\eta_{fan})(\eta_{mot})}$$

From the open system energy equation, the temperature rise across the fan is

$$\Delta T = \frac{\dot{Q}_{fan}}{\rho \dot{V} c_p} \tag{5.33}$$

Example 5.3 Fan Heat Gain
What is the fan heat gain $\dot{Q}_{fan}$ for a fan that has a volumetric flowrate of 9.6 m^3/s at a total pressure of 750 Pa? The fan and motor are in the airstream. The fan efficiency is 60% and the motor efficiency is 88%.

Solution
From Eq. (5.32), the fan heat gain is

$$\dot{Q}_{\text{fan}} = \frac{\dot{V}\Delta P}{(\eta_{\text{fan}})(\eta_{\text{mot}})}$$

$$= \frac{(9.6)(750)}{(0.60)(0.88)} = 13{,}630 \,\text{W}$$

The temperature rise is

$$\Delta T = \frac{\dot{Q}_{\text{fan}}}{\rho \dot{V} c_p}$$

$$= \frac{13630}{(1.18)(9.6)(1005)} = 1.2\,^{\circ}\text{C}$$

Comment: If this fan is to be configured in a blow-through arrangement, the fan heat is added directly to the cooling load, and the dry-bulb temperature entering the coil is increased by 1.2 °C. For a draw through arrangement, the fan heat is absorbed by the supply air and the supply air temperature is increased by 1.2 °C.

5.6   Radiation Heat Transfer

The physical mechanism of thermal radiation heat transfer is energy transfer by electromagnetic waves or photons emitted from one surface and absorbed by another surface. Radiation heat transfer has a major impact on building cooling loads, thermal comfort, and interior heat exchange.

The intensity of thermal radiation, E_b, is a strong function of temperature as it is proportional to the surface temperature to the 4th power, as indicated by the Stefan-Boltzmann blackbody (b) equation, (5.34). For example, at a temperature of 6000 K (sun), $E_b = 7.3 \times 10^7 \,\text{W/m}^2$, and at 300 K (building surfaces) $E_b = 4.7 \times 10^2 \,\text{W/m}^2$.

$$E_b = \sigma T^4 \tag{5.34}$$

The proportionality coefficient σ is the Sigma–Boltzmann constant and has a value of $5.67 \times 10^{-8}\,\text{W/m}^2\text{K}^4$, with the temperature in units of degrees Kelvin (K).

The radiation emitted by a surface has a wavelength distribution. The spectral power has a maximum at a wavelength λ_{max} given by the Wien displacement law,

$$\lambda_{max} = \frac{2898}{T} \tag{5.35}$$

Radiation emitted by the sun at 6000 K has a wavelength peak at about 0.5 microns, in the short-wave visible regime. The radiation emitted by a building surface at a temperature of 300 K has a wavelength peak at about 10 microns, in the long-wave infrared regime.

5.7   Radiation Properties

There are four properties that characterize radiation interaction with a surface. These are emissivity ϵ, absorptivity α, transmissivity τ, and reflectivity ρ. The emissivity ϵ is the ratio of the actual radiation E emitted by a surface to the ideal blackbody radiation E_b,

$$\epsilon = \frac{E}{E_b} \tag{5.36}$$

The actual radiation thus can be expressed as

$$E = \epsilon \sigma T^4 \tag{5.37}$$

Radiation onto a surface can be reflected, absorbed, or transmitted. As shown in Fig. 5.3, the reflectivity ρ is the fraction of radiation reflected, the transmissivity τ is the fraction transmitted, and the absorptivity α is the fraction absorbed. The radiation energy balance is

$$G_{\text{incident}} = G_{\text{absorbed}} + G_{\text{reflected}} + G_{\text{transmitted}}$$
$$1 = \alpha + \rho + \tau \tag{5.38}$$

For an opaque surface, such as a building wall, where $\tau = 0$, Eq. (5.38) becomes

$$\rho = 1 - \alpha \tag{5.39}$$

If the emissivity of a surface is independent of wavelength, the surface is modeled as *grey*. From Kirchoff's law, the absorptivity of a grey surface is equal to its emissivity,

$$\alpha = \epsilon \tag{5.40}$$

However, most surfaces can not be treated as grey when solar radiation is included in the energy balance. As discussed above, the wavelength distribution of solar radiation is very different than the wavelength distribution from a building surface. Therefore, the absorption α_s of short-wave solar radiation by a building surface can be different than the absorption

Fig. 5.3 Radiation energy balance

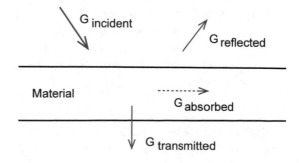

Table 5.3 Emissivity and solar absorptivity of common materials

Material	ϵ	α_s
Ice	0.97	0.3
Concrete	0.90	0.7
Glass	0.94	0.03
Dark paint	0.97	0.98
Red brick	0.93	0.63
Light paint.	0.80	0.21
Snow	0.82	0.28
Galvanized metal	0.28	0.55
Aluminum (shiny)	0.03	0.09

Source Adapted from *2017 ASHRAE Handbook - Fundamentals*, Chap. 4

α and the emission ϵ of infrared radiation. The infrared emissivity ϵ and solar absorptivity α_s are given for various materials in Table 5.3. The emissivity of most building materials is 0.8–0.9, and reflective materials have values around 0.05.

The atmosphere is modeled as a blackbody with an effective sky temperature T_{sky} that is somewhat less than the ground-level air temperature. Radiative emission from the atmosphere is largely from CO_2 and H_2O molecules and is concentrated in the spectral intervals from 5 to 8 microns and above 13 microns. The radiation from the atmosphere to the earth's surface is expressed in the form

$$G_{atm} = \sigma T_{sky}^4 \tag{5.41}$$

The value of T_{sky} depends on atmospheric conditions, ranging from a low of 230 K for a cold clear sky to a high of about 285 K for warm and cloudy conditions.

5.8 Radiation Exchange Between Surfaces

The net radiation transfer between surfaces, such as facing walls in a room, depends on the fraction of radiation emitted by the surface that is absorbed by another surface. To calculate this radiation exchange, we use a radiation network composed of *view factor* and *surface* resistances. The view factor F_{12} is the fraction of radiation that leaves surface 1 that is intercepted by surface 2. Depending on the geometry, view factors vary from 0 to 1, and are independent of surface properties. The surface resistance R_s is a function of the surface emissivity ϵ and is

$$R_s = \frac{1 - \epsilon}{\epsilon} \tag{5.42}$$

For two grey surfaces (1 and 2) facing each other, with surface areas A_1 and A_2, the net rate of radiation heat transfer from surface 1 to surface 2 is given by

$$\dot{Q}_{12} = \frac{\sigma(T_1^4 - T_2^4)}{\frac{1-\epsilon_1}{\epsilon_1 A_1} + \frac{1}{F_{12}A_1} + \frac{1-\epsilon_2}{\epsilon_2 A_2}} \tag{5.43}$$

For the situation where $A_2 >> A_1$ and $F_{12} \simeq 1$ as is the case for radiation from a surface (1) to the atmosphere (2), or a person (1) in a room (2), then Eq. (5.43) can be simplified to

$$\dot{Q}_{12} = \epsilon_1 A_1 \sigma(T_1^4 - T_2^4) \tag{5.44}$$

If the two areas and emissivities are the same, $A = A_1 = A_2$ and $\epsilon = \epsilon_1 = \epsilon_2$, then

$$\dot{Q}_{12} = \frac{\epsilon A \sigma(T_1^4 - T_2^4)}{2 - \epsilon} \tag{5.45}$$

If the two areas are the same, $A = A_1 = A_2$, but with different emissivities, then

$$\dot{Q}_{12} = \frac{A\sigma(T_1^4 - T_2^4)}{\frac{1}{\epsilon_1} + \frac{1}{\epsilon_2} - 1} \tag{5.46}$$

Example 5.4 Net Radiation Heat Transfer
The 100 m² roof of a building is at a temperature of 60 °C (333 K). The roof is coated with white paint and has an emissivity of 0.80. The effective sky temperature is 12 °C (285 K). What is $\dot{Q}_{12}$, the net radiation heat transfer from the roof (1) to the sky (2)?

Solution
Since the roof area (A_1) is much less than the area (A_2) of the atmosphere, and the view factor F_{12} from the roof to the atmosphere is essentially equal to one, we can use Eq. (5.44),

$$\begin{aligned}
\dot{Q}_{12} &= \epsilon_1 A_1 \sigma(T_1^4 - T_2^4) \\
&= (0.80)(100)(5.67 \times 10^{-8})(333^4 - 285^4) \\
&= 2.58 \times 10^{+4} \text{ W}
\end{aligned} \tag{5.47}$$

To include radiation with convection in heat transfer calculations, a radiation heat transfer coefficient $\mathbf{h_r}$ can be derived using Eq. (5.44). Expressing the radiation heat transfer in convection heat transfer form,

$$\dot{Q}_{12} = \epsilon_1 A_1 \sigma(T_1^4 - T_2^4) = \mathbf{h_r} A_1(T_1 - T_2) \tag{5.48}$$

Solving for $\mathbf{h_r}$,

$$\mathbf{h_r} \simeq \epsilon_1 \sigma 4\overline{T}^3 \tag{5.49}$$

where

$$\overline{T} = \frac{T_1 + T_2}{2} \tag{5.50}$$

Table 5.4 Combined convection and radiation thermal resistances of selected building elements

Element	r-value (m^2K/W)
Insulation batt	0.25/cm
Single glazed window	0.16
Double glazed window	0.28
Outside surface (winter)	0.030
Outside surface (summer)	0.044
Inside surface	0.12

Source Adapted from *2017 ASHRAE Handbook—Fundamentals*, Chap. 4

For example, if $T_1 = 100\,°C = 373$ K, $T_2 = 25\,°C = 298$ K, $\epsilon_1 = 0.1$, then $\overline{T} = (373 + 298)/2 = 335.5$ K and

$$
\begin{aligned}
\mathbf{h_r} &\simeq \epsilon_1 \sigma 4 \overline{T}^3 \\
&= (0.1)(5.67 \times 10^{-8})(4)(335.5)^3 \\
&= 0.86 \text{ W/m}^2\text{K}
\end{aligned}
\tag{5.51}
$$

For the typical temperature differences that occur in building applications, the radiation heat transfer coefficient is the same order of magnitude as a natural convection heat transfer coefficient, about 1–5 W/m²K. Combined convection and radiation thermal resistances have been developed for determining heat transfer through various building elements. Selected r-values are given in Table 5.4.

5.9   Multi-mode Heat Transfer

In multi-mode heat transfer, a surface energy balance is used to account for the various modes of heat transfer at the surface. As shown in Fig. 5.4, on an exterior surface of a building, there is radiation from the sun, radiation exchange with the atmosphere, convection to/from the outside air, and thermal conduction through the building surface to/from the interior. At steady state, the thermal energy into a plane surface must equal the thermal energy out of the plane surface.

Example 5.5 Multi-Mode Heat Transfer
What is the roof surface temperature T_r and the net heat transfer per unit area $\dot{Q}/A$ from the roof to the interior of a building when the ambient temperature T_o is 30 °C, the exterior heat transfer coefficient is 25 W/m²C, and the solar flux q_s'' is 750 W/m²? The interior temperature T_i is 22 °C, and the interior heat transfer coefficient is 8 W/m²C. The effective sky temperature $T_{sky} = 280$ K. The roof absorptivity $\alpha_s = 0.21$, long wave emissivity $\epsilon = 0.80$,

Fig. 5.4 Series thermal
resistance network

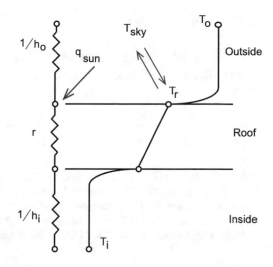

and conduction r-value $r = 2.0 \, \text{m}^2 \text{K/W}$.

Solution

A steady state energy balance on the exterior roof surface is used to find the unknown
temperature T_r. The solar flux $\alpha_s q_s$" absorbed by the surface is equal to the convection to
the ambient environment, conduction to the interior zone at temperature T_i, and net radiation
exchange with the atmosphere,

$$\alpha_s q_s" = \frac{T_r - T_o}{1/\mathbf{h_o}} + \frac{T_r - T_i}{r + 1/\mathbf{h_i}} + \epsilon \sigma \left(T_r^4 - T_{sky}^4 \right)$$

Substituting the numerical values, and solving for T_r,

$$25.47 \, T_r + 4.53 \times 10^{-8} \, T_r^4 = 8.15 \times 10^{+3}$$

or

$$T_r = \frac{8.15 \times 10^{+3} - 4.53 \times 10^{-8} \, T_r^4}{25.47}$$

Assume a guess value of $T_r = 305$, substitute into the right hand side of the above equation,
and iterate to get an updated value of T_r. Within a few iterations we find $T_r = 304.6$, and
so the heat transfer into the interior is $\dot{Q}/A = 4.51 \, \text{W/m}^2$.

As a check, the absorbed solar flux of 157.5 W/m² balances with the three energy paths,
where the net radiation to the atmosphere is 111.7 W/m², convection to the ambient environment is 40 W/m², and the heat transfer into the interior is 4.51 W/m².

5.10   Further Reading

Additional information about heat transfer in air conditioning and refrigeration systems is given in Rohsenow (1998), Bejan (2003), ASHRAE (2017), Cengel (2019), and Bergman (2020).

References

ASHRAE (2017) ASHRAE Handbook - Fundamentals. ASHRAE, Atlanta, Georgia
Bejan A, Kraus A (2003) Heat Transfer Handbook. John Wiley, New York
Bergman T, Lavine A (2020) Fundamentals of Heat and Mass Transfer. John Wiley, New York
Cengel Y, Ghajar A (2019) Heat and Mass Transfer. McGraw Hill, New York
Rohsenow W, Hartnett J, and Cho Y (1998) Handbook of Heat Transfer. McGraw Hill, New York

Cooling Loads in Buildings

6

6.1   Introduction

The cooling loads that the refrigeration and air conditioning systems remove result from the heat gains of the building or conditioned space. If the temperature and humidity are to be held constant, this heat gain needs to be removed at a rate consistent with the rate of heat addition. The rate of thermal energy removal is defined as the cooling load of the building or space. The total cooling load is the sum of both the internal and the external heat gains, each with both sensible and latent components.

The internal heat gain is due to

- People
- Equipment
- Lighting

and the external heat gain is due to

- Ventilation
- Infiltration
- Solar radiation
- Wall Conduction

With commercial and industrial buildings, the internal heat gains are a significant component of the space conditioning load. In fact, many large office buildings require cooling year-round. In these buildings, the largest electricity demand typically is from lighting, followed by refrigeration, computer systems, and ventilation. With residential buildings, the

A. T. Kirkpatrick, *Introduction to Refrigeration and Air Conditioning Systems*,
Synthesis Lectures on Mechanical Engineering,
https://doi.org/10.1007/978-3-031-16776-8_6

external outdoor air introduced to the building through infiltration is a major component of the space conditioning load.

6.2   Sensible and Latent Building Loads

A schematic of a simple forced air cooling system is shown in Fig. 6.1. The system has a cooling coil, i.e., a heat exchanger, to remove the sensible and latent thermal energy from the incoming air stream, and a return flow duct to remove the sensible and latent load from the conditioned space. In the heating season, the cooling coil operates as a heating coil.

The operation of the cooling system is as follows. Air from the outside at state (1) is drawn by a supply fan at a flowrate $\dot{m}_1$, mixed with a portion of the return air, cooled/heated to state (2), and supplied to the space at a flowrate $\dot{m}_2$. Due to the thermal energy load in the space, the space is at state (3), as is the return air. The portion of the return air not recycled is exhausted to the outside, at a flowrate $\dot{m}_1$.

The open system energy equation is used to relate the total load $\dot{L}_t$ to the space mass flow rate and air conditions,

$$\dot{L}_t = \dot{m}_2(h_3 - h_2) \tag{6.1}$$

Similarly, a ventilation (v) load $\dot{L}_v$ is defined using the difference between the outside air and exhaust states,

$$\dot{L}_v = \dot{m}_1(h_1 - h_3) \tag{6.2}$$

The total space load $\dot{L}_t$ is the sum of the sensible (s) and latent (l) load components of the space,

$$\dot{L}_t = \dot{L}_s + \dot{L}_l \tag{6.3}$$

The parameter *Sensible Load Ratio* (SLR) is used to characterize the level of moisture in the space. It is defined as the ratio of the sensible load to the total load,

Fig. 6.1 Cooling System Loads

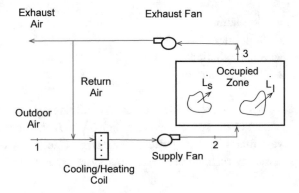

Fig. 6.2 Space Conditioning
Process

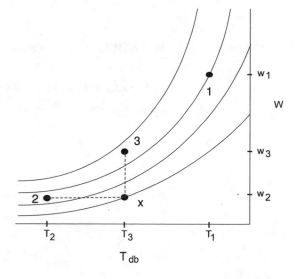

$$\text{SLR} = \frac{\dot{L}_s}{\dot{L}_t} = \frac{\dot{L}_s}{\dot{L}_s + \dot{L}_l} \tag{6.4}$$

A representative space conditioning process is shown on a psychrometric chart in Fig. 6.2 for the cooling system of Fig. 6.1. The outside air at state (1) is relatively hot (T_1) and moist (ω_1). The supply air at state (2) is cold (T_2) and dry (ω_2). The space is at an intermediate state (3). As shown in Fig. 6.2 a virtual state (x) is introduced to separate the sensible and latent components of the cooling load. The state (x) is at the same temperature T_3 of the zone, and the same humidity ratio ω_2 as the supply air. Therefore, expanding Eq. (6.1) into the sensible and latent components,

$$\dot{L}_s = \dot{m}_2(h_x - h_2) = \rho \dot{V}_2 c_{p_a}(T_3 - T_2)$$
$$\dot{L}_l = \dot{m}_2(h_3 - h_x) = \rho \dot{V}_2(h_{fg} + c_{p_v}T_3)(\omega_3 - \omega_2) \tag{6.5}$$

The sensible load $\dot{L}_s$ can be found from a load analysis of the building, with the temperature of the zone and supply air specified. The required outside air $\dot{m}_1$ is determined from indoor air quality considerations.

Example 6.1   Building Cooling Loads
A zone (see Fig. 6.1) has a sensible load of 15 kW and a sensible load ratio (SLR) of 0.8. The supply air leaving the cooling coil is at 8°C and 95% relative humidity. The zone temperature is 24°C. (a) What is the total zone load $\dot{L}_t$, (b) the supply air flowrate $\dot{m}_2$, and (c) the zone humidity ratio ω_3 and relative humidity ϕ_3?

Solution

(a)

$$\dot{L}_l = \frac{\dot{L}_s - \text{SLR}\,\dot{L}_s}{\text{SLR}} = \frac{\dot{L}_s - 0.8\,\dot{L}_s}{0.8} = \frac{0.2(15)}{0.8} = 3.75\,\text{kW}$$

$$\dot{L}_t = \dot{L}_s + \dot{L}_l = 15 + 3.75 = 18.75\,\text{kW}$$

(b) From the psychrometric chart, given T_2, ϕ_2, T_3,

$$h_2 = 23.95$$

$$h_x = 40.25$$

$$\dot{m}_2 = \frac{\dot{L}_s}{(h_x - h_2)} = \frac{15}{40.25 - 23.95} = 0.92\,\text{kg/s}$$

(c)

$$h_3 = \frac{\dot{L}_l}{\dot{m}_2} + h_x = \frac{3.75}{0.92} + 40.25 = 44.32\,\text{kJ/kg}$$

given h_3, T_3, this implies $\omega_3 = 0.0079$ and $\phi_3 = 0.43$.

6.3   Internal Loads

The internal air conditioning loads produced by people depends on their activity level. As discussed previously in Chapter One, the metabolism of humans (and other living creatures) produces heat and moisture that needs to be removed from the conditioned space. The sensible and latent heat produced during different activity levels is given in Table 6.1.

Note that at the light activity levels, the dominant component is the sensible heat production. As the activity level increases, the latent component increases at a faster rate than the sensible, and becomes the largest component.

The equipment heat gain in commercial and industrial buildings is computed on a floor area basis. The equipment heat gain in this type of structure typically varies from 5 to 20 W/m^2, depending on the type of equipment in place. Specialized spaces such as laboratories

Table 6.1 Metabolic rates for different activities

Activity	Sensible (W)	Latent (W)	Total (W)
Rest	70	35	105
Light office work	70	45	115
Heavy office work	80	80	160
Light physical work	80	140	220
Heavy physical work	170	255	425

Source Adapted from *2017 ASHRAE Handbook—Fundamentals*, Chap. 9

Table 6.2 Equipment heat gain

Equipment	(W)
Monitor	1–100
Desktop computer	50–200
Printer	50–500
Copier	50–1500

Source Adapted from *2017 ASHRAE Handbook—Fundamentals*, Chap. 18

require an inventory to determine the equipment heat gain. Data centers have a very large cooling load, of the order of 500–1000 W/m^2, due to their very high density of digital computing devices.

The heat gain from individual office equipment is given in Table 6.2. The lower power consumption values are during standby, and the higher values when operating. A parameter defined as diversity is used to account for the fact that not all of the equipment is in use continually at the same time. A rough estimate of the equipment diversity is 0.50. For a more accurate estimate of the heat gain from a given piece of equipment, a kilowatt meter can be used to measure the power consumption profile.

The lighting load in buildings has been decreasing in recent years, as there have been a number of technological advances in lighting systems, and the adoption of US Federal lighting efficiency standards. Incandescent lights have been replaced by low energy consumption light-emitting diodes (LED). For example, a 7 W LED light produces as much illumination as a 75 W incandescent bulb. Over the last twenty years, the fraction of the electrical load from lighting in commercial buildings has decreased from about 40% to about 15%, however, lighting is still one of the largest end uses of electricity in commercial buildings. A rough estimate of the current (2022) lighting load in a commercial building is about 10–15 W/m^2.

6.4   Weather Data

Weather data for a particular location is needed to estimate the effect of environmental conditions on the cooling load of a building (kW) and the resulting energy consumption (kJ) of the cooling system. For a given location, the cooling load calculation uses the dry and wet bulb temperatures, and the energy consumption calculation uses cooling degree-day data.

Weather data, including dry and wet bulb temperature and cooling degree day data in the US and the world are available in the ASHRAE Handbook of Fundamentals, Chap. 14, *Climatic Design Information* (ASHRAE 2017a). For cooling energy calculation, cooling degree days have been tabulated for balance temperatures of 18.3 °C (65 °F). For determination of cooling loads, the temperature information is given in terms of temperature values that are exceeded by a given percentage of the hours in a year. The percentage typically used for the

Table 6.3 Weather Data

Location	Dry bulb (°C)	Wet bulb (°C)	Difference (°C)	Cooling DD (°C-day)
Phoenix	43.5	20.8	22.7	2576
New Delhi	43.2	22.4	20.8	2971
Denver	34.4	15.9	18.5	401
Cairo	38.2	20.8	17.0	1887
Beijing	35.0	22.0	13.0	871
Berlin	30.0	18.9	11.1	147
Washington, DC	34.8	24.2	10.6	882
Houston	36.0	25.6	10.4	1776
Manila	34.6	26.4	8.2	3727

Source Adapted from *2017 ASHRAE Handbook—Fundamentals*, Chap. 14

selection of cooling equipment, is 0.4%, i.e., 35 hours in a year of 8760 h. Thus, equipment sized using the 0.4% criteria are predicted to have the capacity to meet the cooling load for 99.6% of the time.

The data tables in the ASHRAE Handbook provide the 0.4% dry bulb (db) temperature and the mean coincident wet bulb (wb) temperature, which is the average value of the wet bulb temperature at that dry bulb temperature. For example, in Phoenix, the 0.4% design dry bulb is 43.5 °C and mean coincident wet bulb temperature is 20.8 °C, which means that only 35 h in a year exceed 43.5 °C temperature.

The potential for evaporative cooling of a space depends on the difference between the wet and dry bulb temperatures, as illustrated by the nine representative locations in Table 6.3 below. Hot and dry climates such as Phoenix, Denver, or Cairo have much larger differences between their dry and wet bulb temperatures when compared to more humid climates such as Houston, Washington, DC, or Manila, and so are excellent locations for evaporative cooling systems.

The cooling degree days DD_c are defined in Eq. (6.6) as the integral over a time period t of the positive difference between a balance temperature T_{bal}, and the ambient temperature T_a, and has units °C-day. The plus sign superscript indicates that only positive differences are to be integrated.

$$DD_c = \int_0^t (T_a - T_{bal})^+ \, dt \tag{6.6}$$

The balance temperature T_{bal} is the value of the outdoor temperature where the heat transfer from the space is equal to the internal gains from solar radiation, equipment, and people. It is less than the indoor temperature to provide heat transfer from the space. If the outdoor temperature is greater than the balance temperature, then cooling of the space is required to maintain the indoor temperature.

6.5 Ventilation

Buildings need to maintain a minimum amount of outside airflow to dilute and also remove contaminants and water vapor generated inside the buildings. To remove the contaminants from the space, commercial and industrial buildings use an HVAC system. while residential buildings primarily rely on outside air infiltration. The outside air, if at a warmer temperature and greater humidity than the inside conditions, will add to the overall cooling load. The outside air requirement is typically about $0.010\,\mathrm{m^3/s}$ per person to reduce the CO_2 and water vapor generated by occupants.

ASHRAE Standard 62-2019 *Ventilation for Acceptable Indoor Air Quality* specifies a minimum outdoor airflow for commercial and industrial building zones. The recommended air flowrate $\dot{V}$ depends on the usage patterns of the space, and factors in both the occupant density and the overall size of the space,

$$\dot{V} = R_p P_z + R_a A_z \tag{6.7}$$

where R_p is the outdoor airflow required per person, R_a is the outdoor airflow required per unit floor area, P_z is the number of occupants in the space, and A_z is the zone area. The outdoor airflow requirements for a variety of common indoor spaces is given in Table 6.4.

Example 6.2   Outdoor ventilation air requirement
An office space has a floor area A_z of $1000\,\mathrm{m^2}$ with an occupancy P_z of 70 people. According to ASHRAE Standard 62-2019, what is the outdoor air flowrate requirement?

Solution
Using Eq. (6.7),

$$\dot{V} = R_p P_z + R_a A_z$$
$$= 2.5(70) + 1000(0.3)$$
$$= 475\,\mathrm{L/s} = 0.475\,\mathrm{m^3/s}$$

Table 6.4 ASHRAE Standard 62-2019: Outdoor Airflow Requirements

Category	Air flowrate per person (L/s)	Air flowrate per m^2 (L/s m^2)
Office space	2.5	0.3
Classrooms	5	0.6
Restaurants	7.5	0.9
Supermarket	7.5	0.3
Health club	20	0.3

Source Adapted from *ASHRAE Standard 62-2019*

Common indoor contaminants include carbon dioxide (CO_2), carbon monoxide (CO), radon, volatile organic compounds (VOC), mold, and particulates. Research has indicated that the maximum indoor concentration of CO_2 should be less than 1000 parts per million. Concentrations above this level have been found to be unhealthy.

ASHRAE Standard 62 recommends that the indoor concentration of CO_2 be not more than 700 ppm above the atmospheric concentration. The atmospheric concentration of CO_2 has been rising at a rate of 3-4 ppm per year due to the combustion of fossil fuels. In 1950 the atmospheric concentration of CO_2 was 300 ppm, and in 2020 was 400 ppm.

Carbon monoxide results from incomplete combustion in poorly maintained furnaces and boilers, and is very toxic. The regulated level of CO is 9 ppm maximum during an eight hour period. Volatile organic compounds result from out-gassing from furniture and carpets. VOC compounds include formaldehyde and benzene, and the regulated level of VOC is 1 ppm maximum for an eight hour period.

Ventilation Modeling

Ventilation of buildings can be modeled with an analysis in which the internal sources are assumed to be well-mixed in the zones of the building, and the external concentration of carbon dioxide, $< CO_2 >$ is also considered. The ventilation rate $\dot{V}$ is found using a CO_2 mass flowrate balance. With reference to Fig. 6.3, the exhaust (*exh*) CO_2 flowrate is equal to the outside CO_2 introduced into the zone by the supply (*sup*) air flow plus the CO_2 generated (*gen*) in the zone,

$$\dot{m}_{CO_2,exh} = \dot{m}_{CO_2,sup} + \dot{m}_{CO_2,gen} \tag{6.8}$$

For a constant density, since $\dot{m} = \rho\dot{V}$, the volumetric flowrate balance is

$$\dot{V}_{CO_2,exh} = \dot{V}_{CO_2,sup} + \dot{V}_{CO_2,gen} \tag{6.9}$$

The concentration of carbon dioxide, $< CO_2 >$, is the ratio of the CO_2 flowrate to the air flowrate $\dot{V}$, so dividing Equation (6.9) by $\dot{V}$,

$$< CO_2 >_{exh} = < CO_2 >_{sup} + \frac{\dot{V}_{CO_2,gen}}{\dot{V}} \tag{6.10}$$

Since the CO_2 is assumed to be well mixed, the CO_2 concentration in the zone is the same as the CO_2 concentration in the exhaust, so

$$< CO_2 >_{zone} = < CO_2 >_{sup} + \frac{\dot{V}_{CO_2,gen}}{\dot{V}} \tag{6.11}$$

Solving for the ventilation rate $\dot{V}$,

$$\dot{V} = \frac{\dot{V}_{CO_2,gen}}{< CO_2 >_{zone} - < CO_2 >_{sup}} \tag{6.12}$$

Fig. 6.3 Ventilation Mass Balance

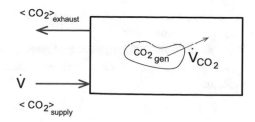

Example 6.3 Zone Ventilation

A sitting person in a room produces CO_2 at a rate of 5.5×10^{-3} L/s. What is the ventilation rate required to maintain the CO_2 concentration in their room at 1000 ppm ? Assume the outdoor air CO_2 concentration is 400 ppm. (1 ppm $= 1 \times 10^{-6}$)

Solution

From Eq. (6.12), the ventilation rate $\dot{V}$ is

$$\dot{V} = \frac{\dot{V}_{CO_2,gen}}{<CO_2>_{zone} - <CO_2>_{sup}}$$

$$= \frac{5.5 \times 10^{-3}}{1.0 \times 10^{-3} - 4.0 \times 10^{-4}}$$

$$= 9.17\,\text{L/s}$$

This ventilation flow rate value compares well with the values given in Table 6.4, which also consider the zone floor area.

6.6   Infiltration

Infiltration is the uncontrolled airflow into/out of a building, typically through small openings in doors, windows, and electrical and plumbing penetrations. Infiltration is responsible for both a sensible and latent energy interaction with the outside environment. This energy interaction can be either a gain or loss, depending on the interior-exterior temperature and humidity difference.

The sensible heat gain (or loss) is

$$\dot{L}_s = \dot{m}\,\Delta h = \rho_a \dot{V} c_{p_a}(T_o - T_i) \tag{6.13}$$

where $\dot{V}$ is the infiltration flowrate (m^3/s), T_o is the outside environment temperature, and T_i is the building inside temperature. The latent heat gain (or loss) is

$$\dot{L}_l = \rho \dot{V}[h_{fg} + c_{p_w}T_o](\omega_o - \omega_i) \tag{6.14}$$

where h_{fg} is the heat of vaporization, ω_o is the exterior humidity ratio, and ω_i is the building interior humidity ratio.

The infiltration flowrate for buildings is also given in the number of **air changes per hour** (ACH), which is defined as the volume flow rate divided by the building or space volume,

$$ACH = \frac{\dot{V}}{V} \tag{6.15}$$

where V is the building interior volume.

Residential buildings rely on infiltration to provide adequate outdoor air for CO_2 and other pollutant removal, so infiltration loss and gain always needs to be included in residential building energy analyses. One rule of thumb is that older residential buildings have on the order of 1 ACH, and newer residential buildings have about 0.3–0.5 ACH. In commercial buildings, the infiltration is relatively small compared to the ventilation airflow, and is usually neglected.

Infiltration Modeling

From the continuity equation, the infiltration flowrate $\dot{V}$ into a building is proportional to an effective infiltration area A_{eff} and a characteristic airflow velocity V,

$$\dot{V} \sim A_{\text{eff}} V \tag{6.16}$$

The effective infiltration area can be measured using a blower door test, in which a fan is used to raise the pressure in the building to a known value for a given flowrate. Bernoulli's equation relates pressure differences ΔP and velocity V,

$$\Delta P \sim V^2 \tag{6.17}$$

Therefore infiltration is proportional to the product of the effective area and the square root of the pressure difference, $P_o - P_i$, between the interior and the exterior of the building,

$$\dot{V} \sim A_{\text{eff}} \Delta P^{1/2} \tag{6.18}$$

The pressure difference between the inside and outside of the building is the sum of two terms, a stack (s) effect and a wind (w) effect.

$$\Delta P = \Delta P_s + \Delta P_w \tag{6.19}$$

The stack effect is due to the inside/outside temperature difference, which creates a density difference and thus a pressure difference that is proportional to the building height. The stack effect is determined with the hydrostatic pressure equation, which gives the decrease in air pressure with height, and the ideal gas equation, which relates temperature and density.

The hydrostatic pressure equation is

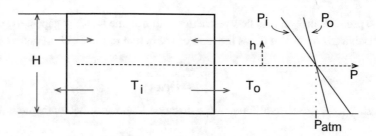

Fig. 6.4 Interior and Exterior Pressure Profiles

$$\frac{dP}{dh} = -\rho g \tag{6.20}$$

If $T_o > T_i$, as is the case in summer, then $\rho_o < \rho_i$, which implies the upward slope or gradient of the inside pressure profile is more negative than the slope of the outside pressure profile, as shown in Fig. 6.4. Note that the two pressure profiles are equal at a height called the neutral plane. If we assume a steady state, the inflow is matched by the outflow, and the neutral height is above the ground level, at about half the height of the building.

Above the neutral plane, for any positive h ($h > 0$), the inside pressure is less then the outside pressure, i.e., $P_o > P_i$, and the airflow is inward from the outside to the inside(infiltration). Below this plane where h is negative ($h < 0$), the outside pressure is less than the inside pressure, $P_i > P_o$, and the airflow is from the inside to the outside of the building (exfiltration). Note that in winter, the pressure gradient and the flow directions are reversed.

Upon integration of the hydrostatic equation, and setting the boundary condition $P_i = P_o = P_{atm}$ at $h = 0$,

$$P_i = P_{atm} - \rho_i g h \quad \text{inside}$$
$$P_o = P_{atm} - \rho_o g h \quad \text{outside} \tag{6.21}$$

The stack pressure difference at height h is

$$\Delta P_s(h) = P_o - P_i = (\rho_i - \rho_o)gh \tag{6.22}$$

If we make the assumption that density differences are due only to temperature differences, known as the Boussinesq approximation, then from the ideal gas law,

$$\frac{\Delta \rho}{\rho} = -\frac{\Delta T}{T} \tag{6.23}$$

so

$$\rho_i - \rho_o = -\rho_i \frac{T_i - T_o}{T_i} \tag{6.24}$$

and the stack pressure difference at height h is

$$\Delta P = P_o - P_i = +\rho_i g h \frac{T_o - T_i}{T_i} \tag{6.25}$$

To take the pressure drop across flow restrictions such as the window gaps into account, we define a discharge coefficient C_d as the ratio of the actual to the ideal pressure difference.

$$C_d = \frac{\Delta P_{actual}}{\Delta P_{ideal}} \tag{6.26}$$

and the actual stack pressure difference at a distance h above or below the neutral plane is

$$P_o - P_i = C_d \rho_i g\, h \frac{T_o - T_i}{T_i} \tag{6.27}$$

Example 6.4   Stack Effect

What are the pressure differences due to the stack effect, a.) at the top, and b.) at the base, of an eight story building in summer with an outside temperature of 35°C (308 K)? The building interior is maintained at 20°C (293 K). Assume $C_d = 0.60$, $\rho_i = 1.18$ kg/m^3, $g = 9.81$ m/s^2, and each story is 3.0 m in height.

Solution

The overall building height H = (8)(3) = 24 m, H/2 = 12 m, $h_{top} = +12$ m, and $h_{base} = -12$ m.

(a) Top of building

$$P_o - P_i = (0.6)(1.18)(9.81)(+12)\frac{35 - 20}{293} = +4.3\,\text{Pa}\quad (\text{infiltration})$$

(b) Base of building

$$P_o - P_i = (0.6)(1.18)(9.81)(-12)\frac{35 - 20}{293} = -4.3\,\text{Pa}\quad (\text{exfiltration})$$

The wind pressure effect is due to the pressure distribution of the wind on the building. The distribution depends on the building orientation and the neighboring features around the building. The pressure is greatest on the windward face and lowest on the sides of the building. Air will infiltrate into the building on the windward face and exfiltrate on the sides. From Bernoulli's equation, the wind stagnation pressure is proportional to the square of the wind velocity.

$$P_{wind} = C_p \frac{\rho V^2}{2} \tag{6.28}$$

The pressure coefficient C_p represents the effect of wind direction on the pressure distribution. Measurements indicate that $C_p \approx 0.6$ on the windward face of a building, $C_p \approx -0.65$ on the low pressure sides, and $C_p \approx -0.3$ on the leeward face.

ASHRAE Infiltration Model

The ASHRAE infiltration model combines the stack and wind coefficients into one equation for the infiltration flowrate $\dot{V}$ (m^3/s),

Table 6.5 Wind Coefficients C_w

Surroundings	Class	One-Story	Two-Story	Three-Story
Unshielded house	1	0.0319	0.0420	0.0494
Rural isolated house	2	0.0246	0.0325	0.0382
House with buildings nearby	3	0.0174	0.0231	0.0271
Suburban house	4	0.0104	0.0137	0.0161
Urban building with houses close by	5	0.0032	0.0042	0.0049

Source Adapted from *2017 ASHRAE Handbook—Fundamentals*, Chap. 16

$$\dot{V} = A_{\text{eff}}(C_s \Delta T + C_w V^2)^{1/2} \tag{6.29}$$

In Eq. (6.29), the term A_{eff} is an effective leakage area (m^2), ΔT is the absolute value of the inside—outside temperature difference (C or K), V_w is the wind speed (m/s), C_s is a stack coefficient (m^2/s^{2}K), and C_w is a dimensionless wind coefficient.

As indicated by Equation (6.30), the stack coefficient C_s is proportional to the number of stories of the building,

$$C_s = 0.015 * (\#\text{Stories}) \tag{6.30}$$

The wind coefficient is given for one to three-story buildings in Table 6.5, where *Class* is the building's surrounding environment listed from one to five, and *Stories* is the number of floors in the building.

A curve fit equation to the values given in the Table 6.5 is

$$C_w = 0.032[1 - 0.225(\text{Class} - 1)][1 + 0.28(\#\text{Stories} - 1)] \tag{6.31}$$

Example 6.5   Infiltration Cooling Load

Estimate (a) the infiltration flowrate in m^3/s and ACH, (b) the sensible and latent cooling loads, and (c) and the cost of summer cooling, due to infiltration into a two story suburban house maintained at 20°C and a humidity ratio of 0.0077 kg$_w$/kg$_a$. The house is at a hot and humid location where the average outdoor summer temperature is 35 °C, the humidity ratio is 0.019 kg$_w$/kg$_a$, and the average wind speed is 4.0 m/s. The house volume is 1200 m^3. A pressurization test indicates that the effective leakage area A_{eff} is 0.20 m^2. The summer cooling season lasts 100 days, the air conditioner coefficient of performance (COP) is 5.0, and the cost of electricity is \$0.12/kWh.

Solution

(a) From Table 6.5, the Building Class is four. Using Eq. (6.30), the stack coefficient C_s is

$$C_s = 0.015(\#\text{Stories}) = 0.015(2) = 0.030$$

The wind coefficient, using the C_w curve-fit equation, Eq. (6.31), is

$$\begin{aligned} C_w &= 0.032[1 - 0.225(\text{Class} - 1)][1 + 0.28(\#\text{Stories} - 1)] \\ &= 0.032[1 - 0.225(4 - 1)][1 + 0.28(2 - 1)] \\ &= 1.33 \times 10^{-2} \end{aligned}$$

The infiltration volumetric flowrate is

$$\begin{aligned} \dot{V} &= A_{\text{eff}}[C_s \Delta T + C_w V^2]^{1/2} \\ &= (0.20)\left[(0.030)(15) + (1.33 \times 10^{-2})4.0^2\right]^{1/2} \\ &= 0.163\,\text{m}^3/\text{s} \end{aligned}$$

The air change per hour (ACH) is

$$\text{ACH} = \frac{\dot{V}}{V} = \frac{0.163}{1200}(3600)\,\text{sec/hr} = 0.488$$

The indoor air is completely changed about every 2 hours, which is typical for a residential building.

(b) The sensible cooling load is

$$\begin{aligned} \dot{L}_s &= \rho \dot{V} c_{p_a} \Delta T \\ &= (1.18)(0.163)(1.0)(15) = 2.88\,\text{kW} \end{aligned}$$

and the latent cooling load is

$$\begin{aligned} \dot{L}_l &= \rho \dot{V}[h_{\text{fg}} + c_{p_w} T_o](\omega_o - \omega_i) \\ &= (1.18)(0.163)[2501.3 + 1.868(35)](0.019 - 0.0077) = 5.58\,\text{kW} \end{aligned}$$

The total infiltration cooling load is therefore

$$\begin{aligned} \dot{L}_t &= \dot{L}_s + \dot{L}_l \\ &= 2.88 + 5.58 = 8.46\,\text{kW} \end{aligned}$$

(c) The electrical power for cooling is

$$\dot{W} = \frac{\dot{L}_t}{\text{COP}} = \frac{8.46}{5.0} = 1.69\,\text{kW}$$

The summer time duration is $\Delta t = (120)(24) = 2880$ hr.
The total electrical energy for cooling use is $W = \dot{W}\Delta t = (1.69)(2880) = 4874\,\text{kWh}$

The cost of cooling is therefore

Cost = (Cost/kWh) (kWh) = ($0.12)(4874) = $585

6.7   Solar Radiation

Solar radiation adds to the air conditioning load of a building either by direct transmission through transparent windows or through absorption on the walls and roof of a building. Since the path of the sun across the sky varies throughout the year, the solar radiation onto a building is not constant, but depends on factors such as building location, orientation, season of the year, time of day, and cloud cover.

The solar radiation entering the earth's atmosphere has a value of about 1367 W/m^2. As the radiation passes through the atmosphere, a portion is scattered, producing diffuse radiation G_d. Other portions are absorbed or reflected back into space, with the remainder, known as beam radiation G_b penetrating onto the earth's surface.

The solar radiation onto a building surface is found as the sum of the direct beam radiation G_b, the diffuse sky radiation G_d, and the ground reflected radiation G_g. The beam radiation onto the exterior surface of a building depends on the orientation of the surface with respect to the sun and is approximately 600–800 W/m^2. The diffuse sky radiation is approximately 100–200 W/m^2. Solar radiation models, see Carroll (1985) and Duffie (1991), such as the ASHRAE clear sky model, have been developed to accurately calculate the beam and diffuse radiation onto the surface of a building at a given time and building location.

As indicated in Fig. 6.5, the sun's position in the sky at a given time is expressed in terms of two angles, the solar zenith angle θ_h and the solar azimuth angle γ. The solar zenith angle θ_h is measured relative to the unit normal vector $\hat{n}$ in the vertical direction. It ranges from 90° when the sun is on the horizon, to 0° if the sun is directly overhead. Negative values correspond to night times. The solar azimuth angle γ is measured on a horizontal plane relative to the southerly direction. It is taken as positive for afternoon hours and negative for morning hours.

The zenith and azimuth angles depend on the location latitude ϕ, solar declination δ, and the hour angle ω. Latitude and longitude values for selected locations are given in Table 6.6. The location latitude ϕ is given in degrees north or south of the equator. The latitude λ is given in degrees relative to Greenwich, England, near London, with the convention used in this text of positive degrees to the west, and negative degrees to east of Greenwich. For example, Denver Colorado is +39.7° north of the equator, and +104.9 degrees east of the 0° meridian.

The solar declination angle δ is a result of the 23.45° tilt of the earth relative to its orbit around the sun, and is given for each day of the year by Eq. (6.32),

$$\delta = 23.45 \sin\left(360 \frac{n_d + 284}{365}\right) \tag{6.32}$$

Fig. 6.5 Solar Zenith θ_h and Azimith γ Angles

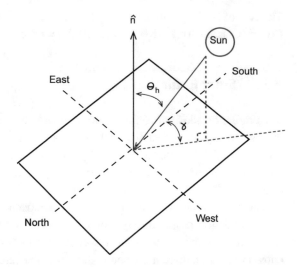

Table 6.6 Latitude ϕ and Longitude λ for Selected Locations

Location	Latitude ϕ (Degrees North)	Longitude λ (Degrees West)
Denver Colorado	+39.7	+104.9
Houston Texas	+29.6	+95.2
Phoenix Arizona	+33.4	+112.0
Washington DC	+38.9	+77.0
Beijing China	+39.9	−116.4
Berlin Germany	+52.5	−13.4
Cairo Egypt	+30.04	−31.2
Manila Philippines	+14.6	−121.0
New Delhi India	+28.6	−77.2

The declination angle δ is in degrees, and n_d is the day number relative to January 1. The declination angle is zero at the equinoxes on March 21 and September 22, has a maximum of 23.45°on June 22, and a minimum of −23.45°on December 22.

The hour angle ω is the angle relative to solar noon when the sun is locally overhead at its highest point in the sky. The hour angle $\omega = 0$°at local solar noon. By convention, the hour angle is negative before solar noon and positive after solar noon. Since the earth takes 24 h or 1440 min. for one rotation of 360°, each degree of solar motion takes about 1440/360 = 4 min., and the hour angle ω relative to noon varies by 360/24 or 15°per hour.

The earth is divided into 24 time zones, each zone about 15°of longitude or one hour in width, starting with 0°at Greenwich, England, near London. Clocks are set for the same reading throughout an entire time zone, defined as the Local Civil Time (LCT) or Standard Time, t_{std} of a selected meridian near the center of the zone. Since the relative motion of the sun is from east to west, solar noon at the eastern edge of a time zone is an hour earlier than solar noon at the western edge.

The time difference Δt in decimal hours between local solar time at a given location and solar time at the standard zone meridian can be found from their longitude difference and the equation of time E_t, as shown in Eq. (6.33). Since the sun moves east to west, locations east of their zone's meridian will have solar noon earlier and the longitude adjustment will be positive. Likewise locations west of their zone meridian will have solar noon later and the longitude adjustment will be negative.

$$\Delta t = \frac{hr}{15\,deg}\,(\lambda_{std} - \lambda_{local}) + E_t \tag{6.33}$$

The equation of time E_t is an equation that takes both the eccentricity of the earth's orbit and the $23.45°$ axial tilt of the earth's ecliptic into account. It is a function of the time of year, with a minimum of about $-0.25\,h$ in February and a maximum of about $+0.25\,h$ in November. It is approximated in decimal hours by Equation (6.34),

$$E_t = 0.0382(0.0075 + 0.1868 \cos \alpha - 3.2077 \sin \alpha - 1.4615 \cos 2\alpha - 4.089 \sin 2\alpha) \tag{6.34}$$

where

$$\alpha = \frac{2\pi}{365}\,(n_d - 1)) \tag{6.35}$$

$$n_d = n^{th} \text{ day of the year}$$

The solar time t_{solar} relative to the local time at a given location is therefore

$$t_{solar} = t_{std} + \Delta t \tag{6.36}$$

and the hour angle ω relative to 12 noon is

$$\omega = \frac{360\,degrees}{24\,hour}(t_{solar} - 12.0) \tag{6.37}$$

The United States has four standard meridians: Eastern Standard Time (EST) at longitude $+75°$ East, Central Standard Time (CST) at longitude $+90°$ East, Mountain Standard Time (MST) at longitude $+105°$ East, and Pacific Standard Time (PST) at longitude $+120°$ East. If Daylight Saving Time is adopted by a location, clocks are advanced one hour ahead of Standard Time in the summer, since there are energy savings by moving an hour of daylight from the morning to the evening.

The solar zenith angle θ_h for a horizontal surface, such as a roof, is given by Eq. (6.38),

$$\cos \theta_h = \sin \delta \sin \phi + \cos \delta \cos \phi \cos \omega \tag{6.38}$$

The solar solar azimuth angle γ is found from Eq. (6.39)

$$\sin \gamma = \frac{\cos \delta \sin \omega}{\sin \theta_h} \tag{6.39}$$

The direct beam component G_h normal to a horizontal roof surface depends on the solar zenith angle θ_h, the angle between the beam radiation and the normal to the wall,

$$G_h = G_b \cos \theta_h \tag{6.40}$$

Example 6.6   Solar Radiation

A flat-roofed building is located in Houston TX. (a) What is the solar zenith angle θ_h and the solar azimuth angle γ at 10:00 am on July 21 ? (b) If the direct beam radiation is 800 W/m², what is the solar flux perpendicular, i. e., normal, to the roof ?

Solution

From Table 6.6 for Houston, the latitude $\phi = +29.6°$and the longitude $\lambda_{loc} = +95.2°$. Houston is in the Central Standard Time Zone with $\lambda_{std} = +90.0°$. July 21 is day $n_d = 202$. (a) The declination angle δ is

$$\delta = 23.45 \sin \left(360 \, \frac{n_d + 284}{365} \right)$$

$$= 23.45 \sin \left(360 \, \frac{202 + 284}{365} \right) = 20.44°$$

The equation of time E_t is

$$\alpha = \frac{2\pi}{365} (n_d - 1)) = 3.46 \, \text{radians}$$

$$E_t = 0.0382(0.0075 + 0.1868 \cos \alpha - 3.2077 \sin \alpha - 1.4615 \cos 2\alpha - 4.089 \sin 2\alpha)$$

$$= -0.106 \, \text{hr}$$

The time difference Δt is

$$\Delta t = \frac{\text{hr}}{15 \deg} (\lambda_{std} - \lambda_{local}) + E_{time}$$

$$= \frac{\text{hr}}{15 \deg} (+90 - (+95.2)) + (-0.106) = -0.453 \, \text{hr}$$

Since Houston is west of the zone meridian, the longitude correction is negative.
The solar time relative to the local time at a given location is therefore

$$t_{solar} = t_{std} + \Delta t = 10.0 - 0.453 = 9.55 \, \text{hr}$$

and the hour angle ω relative to 12 noon is

$$\omega = \frac{360}{24}(t_{solar} - 12.0) = \frac{360}{24}(9.55 - 12) = -36.8°$$

The solar zenith angle θ_h is

$$\cos \theta_h = \sin \delta \sin \phi + \cos \delta \cos \phi \cos \omega$$
$$= \sin(20.44) \sin(29.6) + \cos(20.44) \cos(29.6) \cos(-36.8) = 0.825$$
$$\theta_h = \cos^{-1} 0.825 = +34.4°$$

The solar solar azimuth angle γ is

$$\sin \gamma = \frac{\cos \delta \sin \omega}{\sin \theta_h}$$
$$= \frac{\cos(20.44) \sin(-36.8)}{\sin(34.4)} = -0.99$$
$$\gamma = \sin^{-1}(-0.99) = -83.5°$$

The azimuth angle is toward the east, as it is before solar noon.

(b) The direct beam component G_h normal to the roof surface is

$$G_h = G_b \cos \theta_h = (800) \cos(34.4) = 660 \text{ W/m}^2$$

6.8   Wall Conduction

Determination of contribution of the building wall heat transfer to the cooling load is complicated due to the thermal capacitance of the building walls. The thermal capacitance introduces a time delay into the thermal energy penetration rate. For example, as the outside heat transfer into the building increases during the day, thermal energy is stored in the walls and is subsequently released at a later time as the building walls cool. There are two main techniques used by building energy engineers for determination of the wall conduction, Heat Balance and Radiant Time Series.

The Heat Balance method, Pedersen et al. (1997), uses an energy balance model including lumped capacitance. The wall heat conduction and surface convection equations are written in a finite difference formulation, where the heat transfer into and out of a given finite difference node is calculated from the nodal temperature difference divided by the thermal resistance, with a similar treatment for the thermal storage terms. The conduction heat transfer through the walls is found from the solution of the energy balance equations for the interior zones and the surfaces of each wall, roof, and floor. The outside surface energy balance includes absorbed short wave solar radiation, convection from the surrounding air, and long wavelength radiation from the ground and sky, all of which contribute to conduction heat transfer through the exterior wall into the building interior.

The Radiant Time Series (RTS) method, Seem et al. (1989), is a transfer function approach. The transfer function coefficients quantify the contribution of each building element to the overall cooling load. The heat transfer rate from the surfaces of an element is represented as a time series of the current and previous temperatures and heat fluxes multiplied by the transfer function coefficients. The coefficients are tabulated in the ASHRAE Handbook for a variety of wall and roof types. However, it should be noted that the coefficients do not directly relate to the wall physical properties such as thermal conductivity and density.

6.9   Further Reading

Two textbooks that provide additional information about the the analysis and modeling of building cooling systems are Kreider et al. (2010) and Mitchell and Braun (2013).

References

ASHRAE (2017a) ASHRAE Handbook – Fundamentals, Chapter 14-Climatic Design Information. ASHRAE, Atlanta, GA

ASHRAE (2017b) ASHRAE Handbook – Fundamentals, Chapter 17 –Cooling and Heating Load Calculations. ASHRAE, Atlanta, GA

Carroll J (1985) Global transmissivity and diffuse fraction of solar radiation for clear and cloudy skies as measured and as predicted by bulk transmissivity models. Solar Energy 35(2):105-118

Duffie J, Beckman W (1991) Solar Engineering of Thermal Processes. John Wiley, New York

Kreider J, Curtiss P, Rabl A (2010) Heating and Cooling of Buildings. CRC Press, Boca Raton, Florida

Mitchell J, Braun J (2013) Principles of Heating, Ventilation, and Air Conditioning in Buildings. John Wiley, New York

Pedersen C, Fisher D, Liesen (1997) Development of a heat balance procedure for calculating cooling loads. ASHRAE Transactions 103(2):459-468

Seem J, Klein W, Beckman W, Mitchell J (1989) Comprehensive room transfer functions for efficient calculation of transient heat transfer in buildings. ASME J Heat Transfer 111(5):264-273

Cooling Equipment

7

7.1 Introduction

In this chapter we apply the heat transfer and fluid mechanics analyses presented in previous chapters to determine the actual size and the thermal performance of the system components needed to meet a required cooling load. For example, in a vapor compression system, there are four basic mechanical components: a condenser, evaporator, compressor, and an expansion device. Each of these components needs to be sized and then assembled into a cooling system which will operate in a cycle to meet a given design cooling load.

For each of these system components, the engineer has a wide range of alternative configurations to choose from. Many pre-built cooling system packaged units are commercially available, in which the various system components have been selected, balanced, assembled, and charged with refrigerant for delivery of a working unit to a site. This is a simpler option relative to selection of the individual components, however, the match with the design load may not be exact.

The unified package of vapor compression components (compressor, evaporator, condenser, and expansion device) used to produce chilled water for air conditioning of commercial buildings is defined as a chiller. A split package air conditioner is defined as a cooling system with the evaporator and condenser in separate locations.

7.2 Heat Exchangers

We begin with heat exchangers, since they are common to all of the thermodynamic cooling cycles examined. Heat exchangers are used in cooling systems to transfer thermal energy from a *hot* fluid to a *cold* fluid. For example, in a vapor compression system the working fluid in the evaporator absorbs thermal energy from the object or space to be cooled through

© The Author(s), under exclusive license to Springer Nature Switzerland AG 2023
A. T. Kirkpatrick, *Introduction to Refrigeration and Air Conditioning Systems*,
Synthesis Lectures on Mechanical Engineering,
https://doi.org/10.1007/978-3-031-16776-8_7

evaporation from a liquid to a vapor. In the condenser, the working fluid transfers thermal energy to the environment through condensation from a vapor to a liquid.

In refrigeration systems where the temperatures are below 0 °C, and secondary refrigerant loops are needed, brine solutions are used to lower the freezing point of the working fluid. A brine is a solution of water and an inorganic salt, either sodium chloride or calcium chloride. Sodium chloride and propylene glycol are compatible with food materials. However, calcium chloride and ethylene glycol are not recommended for food related applications.

There are many types of heat exchanger geometries used in cooling systems, the three most common geometries used in cooling systems being counterflow, crossflow, and shell and tube. Counterflow heat exchangers are used when both sides of the heat exchanger are liquids. A schematic of a counterflow configuration is shown in Fig. 7.1. The hot and cold liquids enter the heat exchanger from opposite sides so that depending on the flow rate ratio, it is possible for the cold side outlet temperature to be greater than the hot side outlet temperature, as illustrated in Fig. 7.2. With a tube-in-tube counterflow heat exchanger, water flows in the inner tube and the refrigerant flows in the opposite direction in the annular space between the inner and outer tubes.

Crossflow heat exchangers are used when one or more of the fluids is a gas, for example, a cooling coil in an air conditioning system. In a crossflow heat exchanger, as shown in Fig. 7.3, the liquid flow is through finned tubes which are arranged perpendicular to the gas flow. Fins are placed on the exterior of the tubes to increase the gas side heat transfer coefficient. The cooling coils in air conditioning systems are finned tubes in cross flow, with chilled water or refrigerant flowing inside the tubes, and air flowing outside the tubes.

Fig. 7.1 Schematic of counterflow heat exchanger

Fig. 7.2 Temperature profile in a counterflow heat exchanger

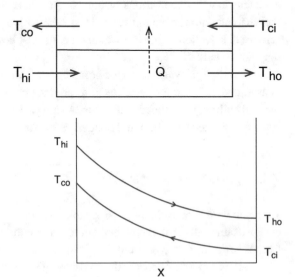

Fig. 7.3 Schematic of
crossflow heat exchanger

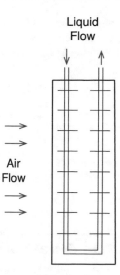

Coils are composed of finned tubes running across the width of the coil, arranged in rows. The coils typically have 4–6 rows with fin spacings of about 3–5 fins per cm. The number of continuous flow paths from the fluid supply to return is determined by the arrangement of tube connections at the end of the rows. Each portion of the fluid flow makes a number of passes across the width of the coil. The air-side pressure drop is typically 125–250 Pa, and the fluid side pressure head drop is from 2 to 6 m.

Selection of the number of rows and fins, and the trade-offs between air and fluid pressure drop are found from an economic analysis of fan and pump operating costs and equipment first cost. For example, heat transfer performance can be improved by providing more closely spaced fins, which increases air pressure drop. The same performance could be achieved with fewer fins but with additional rows, increasing the fluid pressure drop.

Performance data for cooling coils is obtained from manufacturer's catalogs. The catalogs provide data on coil pressure drop, coil heat and mass transfer, and leaving dry and wet bulb temperature for given entering air dry and wet bulb temperatures. Many coiling coil manufacturers provide performance ratings certified under ARI Standard 410–2001. This standard provides a method for extending laboratory data to operating conditions and coil configurations other than those tested. The Standard covers performance ratings to a minimum fluid velocity of 0.3 m/s, and air face velocity of 1–4 m/s.

Shell and tube heat exchangers are used in large condensers and evaporators. A shell and tube heat exchanger has tubes arranged in a symmetric pattern inside a cylindrical shell. A schematic of a shell and tube condenser with one refrigerant shell pass and two water passes is shown in Fig. 7.4. Manifolds in the end plates direct the water flow into the tubes, and baffles in the shell direct the shell flow past the tubes.

Fig. 7.4 Schematic of a shell
and tube condenser

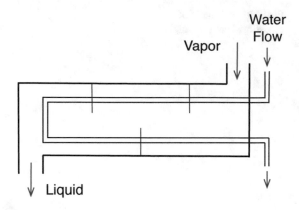

7.3   Heat Transfer in Heat Exchangers

The rate of thermal energy transport in the heat exchangers of cooling systems is a function of
the heat transfer mechanism, fluid properties, and the heat exchanger size and geometry. The
primary mechanisms of sensible heat transfer in condensers and evaporators are conduction
and convection, with radiation playing an insignificant role. The heat transfer coefficient
depends on the type of convection. Natural convection results from buoyancy forces created
by the surface being at a different temperature than the fluid, forced convection results from
an external pressure difference created by a pump, fan, or wind. Boiling heat transfer results
from a change of phase from a liquid to a vapor when a liquid contacts a surface with a
temperature greater than the saturation temperature of the liquid. Condensation heat transfer
is the reverse of boiling, and results from a change of phase when a vapor contacts a surface
with a temperature less than the saturation temperature of the vapor.

There are two main analysis techniques for heat exchangers. They are designated as
the Log Mean ΔT method, and the Effectiveness-Number of Transfer Units ($\epsilon - NTU$)
method. Both techniques use the energy equation and Newton's convection equation to relate
the heat transfer to the mass flow rates and temperatures of the hot and cold fluids.

The $\epsilon - NTU$ method is more widely used for heat exchanger design and selection,
since it only requires the inlet temperatures of the hot and cold fluids as inputs. The heat
exchanger effectiveness ϵ is defined as the ratio of the actual heat transfer to the maximum
possible heat transfer.

$$\epsilon = \frac{\dot{Q}_{act}}{\dot{Q}_{max}} \tag{7.1}$$

The Number of Transfer Units (NTU), a dimensionless parameter, is defined as

$$NTU = \frac{UA}{(\dot{m}c)_{min}} \tag{7.2}$$

where U is the overall heat transfer coefficient, and A is the heat exchanger area. The maximum heat transfer is given by the product of the minimum heat capacity rate $(\dot{m}c)_{min}$ and the maximum temperature difference between the hot fluid inlet (hi) and the cold fluid inlet (ci)

$$\dot{Q}_{max} = (\dot{m}c)_{min}(T_{hi} - T_{ci}) \qquad (7.3)$$

where $\dot{m}$ is a fluid mass flowrate and c is a fluid specific heat. The actual heat transfer is the thermal energy transferred from the hot side to the cold side.

$$\dot{Q}_{act} = (\dot{m}c)_h(T_{hi} - T_{ho}) = (\dot{m}c)_c(T_{co} - T_{ci}) \qquad (7.4)$$

For condensers and evaporators, if we assume the superheating and subcooling heat transfer is small relative to the phase change heat transfer, the actual heat heat transfer is

$$\dot{Q}_{act} = \dot{m}\,h_{fg} \qquad (7.5)$$

Given ϵ, T_{hi}, and T_{ci}, the actual heat transfer rate can be expressed as

$$\dot{Q}_{act} = \epsilon\,(\dot{m}c)_{min}(T_{hi} - T_{ci}) \qquad (7.6)$$

It can be shown that the effectiveness for a heat exchanger is

$$\epsilon = f\left(NTU, \frac{(\dot{m}c)_{min}}{(\dot{m}c)_{max}}\right) \qquad (7.7)$$

Furthermore, for condensers and evaporators with a constant temperature on the condensing or evaporating side, the effectiveness is only a function of the parameter NTU. The relationship is

$$\epsilon = 1 - e^{-NTU} \qquad (7.8)$$

Example 7.1 Heat Exchanger Analysis

An air-cooled condenser has a finned tube crossflow geometry. The condenser needs to be sized to transfer $\dot{Q}_{act} = 5.0$ kW of thermal energy. The working refrigerant is R-134a at a condensing temperature of 44 C. The overall heat transfer coefficient U is 500 W/m²-K, and the diameter D of the thin walled finned tube is 30 mm. The air, initially at 20 C, has a temperature rise of 10°C, and has a specific heat of 1000 J/kg-K. The latent heat of the refrigerant R-134a is 1.80×10^5 J/kg. What are the refrigerant (r) and air (a) flowrates in the condenser ? What is the required condenser tube length L ? Neglect the superheating and subcooling portions of the condensation process.

Solution

From Eq. (7.5), the refrigerant flowrate is

$$\dot{m}_r = \frac{\dot{Q}_{act}}{h_{fg}} = \frac{5.0 \times 10^3}{1.80 \times 10^5} = 0.028\,\text{kg/s}$$

The air flowrate is

$$\dot{m}_a = \frac{\dot{Q}_{act}}{c_p \,\Delta T} = \frac{5.0 \times 10^3}{(1000)(10)} = 0.50 \,\text{kg/s}$$

The minimum heat capacity rate is on the air side,

$$\dot{Q}_{max} = (\dot{m}c)_{min}(T_{hi} - T_{ci}) = (0.50)(1000)(44 - 20) = 12000 \,\text{W}$$

The effectiveness is

$$\epsilon = \frac{\dot{Q}_{act}}{\dot{Q}_{max}} = \frac{5000}{12000} = 0.416$$

The NTU value is

$$NTU = -\ln(1 - \epsilon) = -\ln(1 - 0.416) = 0.539$$

and the tube area A is

$$A = \frac{(NTU)(\dot{m}c)_{min}}{U} = \frac{(0.539)(0.5)(1000)}{500} = 0.53 \,\text{m}^2$$

Therefore the tube length L is

$$L = \frac{A}{\pi D} = \frac{0.53}{\pi \,(.03)} = 5.7 \,\text{m}$$

Due to its length, the tube will need to be bent into a U shape with multiple passes.

7.4   Condensers

The condenser operates on the high temperature, high pressure side of the cooling cycle. Since the condenser transfers the thermal energy extracted from the refrigerated object and from the compressor to the environment, the working fluid needs to be at a greater temperature than the environment. From the first law applied to the cooling cycle, the thermal energy rejected to the environment is the sum of the evaporator heat transfer $\dot{Q}_l$ and the pump or compressor work $\dot{W}$.

$$\dot{Q}_h = \dot{Q}_l + \dot{W} \tag{7.9}$$

The working fluid enters the condenser as a superheated vapor and leaves as a condensed liquid. Most of the heat transfer is isothermal, however, there are usually a few degrees of desuperheating of the entering vapor and subcooling of the exiting liquid flow.

The thermal energy to be transferred from the condenser is transferred to the atmosphere directly using an air cooled condenser or indirectly using a cooling tower. The choice of condenser depends primarily on the application and capacity required. The three major types of condensers are air-cooled, liquid-cooled, and evaporative-cooled. The heat transfer coefficient on the condensing side of the heat exchanger is of the order of 10,000 W/m²-K, so the major resistance to heat transfer is on the environment air/water side.

Air-cooled condensers are used in domestic refrigeration and air conditioning systems with capacities less than 500 kW. They are cross flow heat exchangers that use a finned tube geometry. The refrigerant vapor enters at the top, and the condensing liquid flows by gravity to the exit at the bottom.

For applications requiring capacities of about 200–1000 W, the heat transfer to the air can be provided by natural convection. Example applications are domestic refrigerators, freezers, small commercial, and laboratory equipment. Natural convection condensers are constructed with bare or finned tubing, or tubes attached to thin panels. The condenser is located a few centimeters off the back of the unit to establish a natural convection chimney effect. The refrigerant enters the top of the condenser as a vapor and condenses to a liquid as it flows downward under the action of gravity. As the air outside the tube is heated by the condenser, it rises due to its increased buoyancy and is replaced by cooler air entering at the bottom of the condenser.

Single stage axial flow fans are used in rooftop air-cooled condensers for buildings with required cooling capacities greater than 5–10 kW and are shown schematically in Fig. 7.5. The fans provide forced convection air flow either vertically or horizontally over a finned tube geometry. Fins are used on the air side of the condenser to reduce the thermal resistance and increase the heat transfer rate. As shown in Fig. 7.5, multiple fans are used to reduce noise and modulate load. The temperature rise of the ambient air stream is designed to be of the order of 10 K or less to maintain an acceptable temperature difference across the condenser.

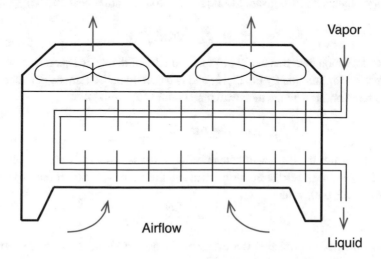

Fig. 7.5 Schematic of forced convection air-cooled condenser

Cooling systems with a required capacity greater than 100 kW generally also use liquid cooling, since water has a much greater heat capacity and density than air. Smaller condenser configurations use a double walled pipe operated with the refrigerant being in the annulus. Larger condenser configurations use a shell and tube, with water in the tube, and the refrigerant vapor condensing to a liquid in the shell, as illustrated in Fig. 7.4 presented in the previous section. Hot refrigerant vapor enters at the top of the shell, condensing when it comes into contact with the cold surface of the water tubes or coils. The condensed refrigerant drips off the coils, collecting in the bottom of the shell, which also serves as a receiver tank.

In a liquid cooled condenser, the liquid side temperature rise $(T_2 - T_1)$ is typically of the order of 5–10 K, and the mass flowrate $\dot{m}_w$ of the water is found from an energy balance on the liquid side,

$$\dot{m}_w = \frac{\dot{Q}_h}{c_p\,(T_2 - T_1)} \tag{7.10}$$

7.4.1 Condensation Heat Transfer

When the superheated refrigerant vapor comes into contact with the cold walls of a condenser, the vapor condenses into droplets and forms a liquid film on the tube wall. The liquid film flows downward and pools at the bottom of the tube, as gravity and a shear force from the vapor flow pushes the liquid pool downstream along the tube. The vapor mass fraction, i.e., quality, decreases from 1 to 0 from the entrance to the exit of the condenser. As a result of the phase change h'_{fg} from vapor to liquid, the condensation heat transfer $\dot{Q}_{cond}$ can be expressed as

$$\dot{Q}_{cond} = \dot{m}_v h'_{fg} \tag{7.11}$$

The condensation heat transfer $\dot{Q}_{cond}$ inside a tube of diameter D can also be expressed in a convection formulation by Eq. (7.12) and is based on the difference between the vapor saturation temperature and the tube wall temperature $(T_{sat} - T_s)$.

$$\dot{Q}_{cond} = \bar{h}_D A (T_{sat} - T_s) \tag{7.12}$$

The heat transfer coefficient in condensation processes depends on the saturation and tube wall temperatures, refrigerant properties and the vapor velocity. If the vapor Reynolds number $Re < 35,000$, where

$$Re = \frac{4\,\dot{m}}{\pi\,D\mu_v} \tag{7.13}$$

then a recommended correlation (Dobson and Chato 1998) for the average condensation heat transfer coefficient inside a horizontal tube is

$$\bar{h}_D = 0.555 \left[\frac{g\rho_l(\rho_l - \rho_v)k_l^3 h'_{fg}}{\mu_l(T_{sat} - T_s)D} \right]^{1/4} \tag{7.14}$$

The term h'_{fg} is the corrected latent heat accounting for the subcooling of the condensed vapor.

$$h'_{fg} = h_{fg} + 0.375 c_{pl}(T_{sat} - T_s) \tag{7.15}$$

The term ρ_l is the liquid density, ρ_v is the vapor density, k_l is the liquid thermal conductivity, c_{pl} is the liquid specific heat, and μ_l is the liquid dynamic viscosity. The liquid properties are evaluated at the mean liquid film temperature, and the vapor properties are evaluated at the saturation temperature.

$$T_f = (T_{sat} + T_s)/2 \tag{7.16}$$

Example 7.2 Condensation Heat Transfer
A horizontal tube air cooled condenser uses refrigerant R-134a. The refrigerant enters the condenser tube at a flowrate of 0.01 kg/s as a saturated vapor at 44 °C and exits as a saturated liquid. The tube is 30 mm in diameter with a constant wall temperature of 24 °C. How long is the condenser tube?

Solution
The mean liquid film temperature is

$$T_f = (T_{sat} + T_s)/2 = (44 + 24)/2 = 34°C$$

The properties (ASHRAE 2017) of the saturated vapor at 44 °C are ρ_v = 56.0 kg/m³, h_{fg} = 1.587×10⁵ J/kg, and μ_v = 12.8 Pa-s. The properties of the liquid condensate at the mean film temperature of 34 °C are ρ_l = 1172 kg/m³, k_l = 0.0773 W/m-K, c_{pl} = 1466 J/kg-K, and μ_l = 1.766 ×10⁻⁴ Pa-s. The corrected heat of vaporization is

$$h'_{fg} = h_{fg} + 0.375 c_{pl}(T_{sat} - T_s) = 1.587 \times 10^5 + 0.375(1466)(20) = 1.70 \times 10^5 \text{ J/kg}$$

The condensation heat transfer coefficient $\bar{h}_D$ is

$$\bar{h}_D = 0.555 \left[\frac{g\rho_l(\rho_l - \rho_v)k_l^3 h'_{fg}}{\mu_l(T_{sat} - T_s)D} \right]^{1/4}$$

$$= 0.555 \left[\frac{9.8(1172)(1172 - 56)(0.0773)^3(1.70 \times 10^5)}{1.766 \times 10^{-4}(44 - 24)(0.03)} \right]^{1/4} = 974 \text{ W/m}^2\text{-K}$$

Since

$$\dot{Q}_{cond} = \dot{m}_v h'_{fg} = \bar{h}_D \pi DL(T_{sat} - T_s)$$

then

$$L = \frac{\dot{m}_v h'_{fg}}{\bar{h}_D \pi D (T_{sat} - T_s)} = \frac{0.01(1.70 \times 10^5)}{974\,\pi\,(0.03)(20)} = 0.93\,\text{m}$$

7.4.2   Evaporative Condensers

Evaporative condensers are a combination of an air-cooled condenser and a cooling tower. In addition to a cooling air flow, they use a water spray onto the outside of the condenser tube bank to cool and condense the refrigerant inside the tubes. The heat transferred from the condensing refrigerant evaporates the water film on the condenser tubes and increases the temperature of the air stream.

Water is pumped from a sump at the bottom of the unit to nozzles located at the top and then sprayed as a mist onto the condenser tubes. The water that does not evaporate falls under gravity back to the sump. Thus both the temperature and the humidity of the air stream increase. The rate of heat transfer depends on both the dry bulb and the wet bulb temperature of the inlet air stream. As the inlet air temperature decreases, the the condenser effectiveness increases.

7.5   Evaporators

The evaporator operates on the low temperature, low pressure side of the cooling cycle, absorbing thermal energy from the space or object to be cooled. Since it is in thermal contact with the cooling load, the refrigerant enters the evaporator as a two phase liquid, and leaves as a vapor. The working fluid on the hot side of the evaporator is usually air or water. There are many configurations used for evaporators, depending on the application and capacity required.

For air cooled evaporators, the most common configuration is direct expansion (DX) using finned tube heat exchangers in which the refrigerant flow in the tubes is evaporated by the heat transfer from the hot air stream flowing around the tubes. The tube geometries used are bare tube, finned tube, and plate-type. One configuration of plate type evaporators uses two sheets of corrugated metal that are bonded together to form channels for the refrigerant flow. Another configuration widely used in refrigerator trucks consists of tubes sandwiched between two plates. Finned tube evaporators are made by bonding metal plates or fins to a bare tube to increase the heat transfer surface area. The fin spacing varies from about 2 mm on a domestic air conditioning unit to 10 mm for a cold room heat exchanger. The finned tubes are horizontal, with vertical fins to allow defrosting and drainage of condensed moisture in air flow conditions below the dew or frost points.

The air circulating on the 'hot' side flows by either natural or forced convection. A cross-flow geometry is used in which the air flow is perpendicular across the tubes. The flow arrangements on the refrigerant side are of a series or parallel configuration. Small capacity evaporators such as domestic refrigerators use a single tube, and larger capacity evaporators use a parallel flow geometry with multiple tubes to reduce the pressure drop across the evaporator.

For liquid cooled evaporators, a common configuration is a shell and tube heat exchanger. The tubes can be straight, U-shaped, or coiled. For cooling loads less than 500 kW, the refrigerant flow is generally inside the tubes, termed dry expansion, and the water or brine is in the shell. The refrigerant evaporation process is convective boiling inside the tubes. To prevent liquid droplets from entering the compressor, dry expansion evaporators will superheat the refrigerant vapor a few degrees above saturation. The overall heat transfer coefficient U is of the order of 500–1200 W/m²-K for dry expansion evaporators. This configuration is also used for ice manufacture, where ice forms on the outside of the evaporator coils.

For larger cooling loads above 500 kW, the refrigerant is generally on the shell side, with a circulating liquid originating from the cooling load passing through the tubes, as shown in the schematic in Fig. 7.6. This arrangement is termed a flooded evaporator, since the refrigerant liquid level is maintained high enough so the tubes are always covered by liquid refrigerant. The tubes at top of the shell are removed to reduce entrainment of the liquid refrigerant by the boiling vapor. The liquid level can be maintained constant using a float valve external to the evaporator. The overall heat transfer coefficient U is of the order of 700–1200 W/m²-K for flooded evaporators.

In a flooded evaporator, the water flowing in the tubes will decrease in temperature, typically 5–10 °C, so the wall temperature is not constant. Consequently, the resulting heat transfer is a combination of nucleate boiling, natural convection, and evaporation from the surface of the pool at the top of the shell. The exterior surface of the tubes need to be about 10 °C above the saturation temperature for nucleate boiling to occur. In nucleate pool boiling, vapor bubbles form on the tube surface, grow in size, and detach when their

Fig. 7.6 Flooded shell and tube evaporator

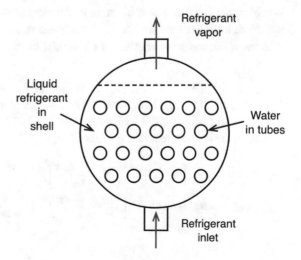

buoyancy exceeds the surface tension restoring force. The bubble detachment from the tube agitates the pool of liquid refrigerant and provides a mechanism for a microlayer of colder liquid refrigerant to begin forming a new bubble on the surface of the tube.

7.6   Boiling Heat Transfer

The nucleate boiling heat transfer coefficient $\bar{h}$ on the exterior surface of the tubes depends on a number of factors, including the tube wall - fluid saturation temperature difference, $T_s - T_{sat}$, surface finish, and fluid properties. The heat transfer correlations are of the form

$$\bar{h} = C\,(T_s - T_{sat})^n \tag{7.17}$$

Since the heat transfer coefficient is a function of the temperature difference to some power, the correlations are expressed directly in terms of the heat flux $\dot{Q}/A$. A widely used correlation (Rohsenow et al. 1998) is given in Eq. (7.18). Note the correlation is independent of the tube diameter and length.

$$\frac{\dot{Q}}{A} = \mu_l\,h_{fg}\left[\frac{g(\rho_l - \rho_v)}{\sigma}\right]^{1/2}\left[\frac{c_{pl}\,(T_s - T_{sat})}{C_s\,h_{fg}}\right]^3\left(\frac{c_{pl}\mu_l}{k_l}\right)^{-3} \tag{7.18}$$

The term ρ_l is the liquid density, ρ_v is the vapor density, k_l is the liquid thermal conductivity, c_{pl} is the liquid specific heat, μ_l is the liquid kinematic viscosity, σ is the surface tension, and C_s is a surface finish coefficient. The liquid and vapor properties are evaluated at the saturation temperature.

Example 7.3 Nucleate Boiling Heat Transfer
The refrigerant in the shell of a horizontal tube flooded evaporator for a commercial freezer is refrigerant R-134a. The refrigerant is at a saturation temperature of $-20\,^\circ$C, and the tube wall is at a temperature of $-5\,^\circ$C. What is the nucleate boiling heat flux from the tubes ? Assume a surface finish coefficient $C_s = 0.0128$.

Solution
The properties (ASHRAE, 2016) of the saturated vapor at $-20\,^\circ$C are $\rho_v = 6.8$ kg/m^3 and $h_{fg} = 212.9\times10^3$ J/kg. The properties of the saturated liquid are $\rho_l = 1358.3$ kg/m^3, $k_l = 0.1011$ W/m-K, $c_{pl} = 1293$ J/kg-K, $\mu_l = 353\times10^{-6}$ Pa-s, and $\sigma = 14.51\times10^{-3}$ N/m. The nucleate boiling heat flux $\dot{Q}/A$ is

$$\frac{\dot{Q}}{A} = \mu_l\,h_{fg}\left[\frac{g(\rho_l - \rho_v)}{\sigma}\right]^{1/2}\left[\frac{c_{pl}\,(T_s - T_{sat})}{C_s\,h_{fg}}\right]^3\left(\frac{c_{pl}\mu_l}{k_l}\right)^{-3}$$

$$= 0.000353(212.9\times10^3)\left[\frac{9.8(1358.3 - 6.8)}{14.51\times10^{-3}}\right]^{1/2}\left[\frac{1293\,(15)}{0.0128(212.9\times10^3)}\right]^3\left(\frac{1293(0.000353)}{0.1011}\right)^{-3}$$

$$= 2.81\times10^5 \text{ W/m}^2$$

7.7 Cooling Towers

With large (>500 kW) vapor compression and absorption cooling systems, the thermal energy absorbed by the water on the cooling side of the condenser is transferred to a cooling tower for rejection to the atmosphere. In the cooling tower, thermal energy is transferred from the water to the atmosphere by evaporation of a portion of the water in the form of droplets.

The mechanical draft cooling tower is used in most large cooling systems. This is a configuration in which water is sprayed downward into an upward flowing air stream driven by fans. Fans are located at the top of the tower in a induced-draft cooling tower, as shown in Fig. 7.7, and at the bottom of the tower in a forced-draft tower. Induced-draft cooling towers are preferred, as forced-draft towers can have issues with freezing and short circuiting. In a crossflow mechanical draft cooling tower, the airflow is horizontal, and thus perpendicular to the downward flowing water stream. The heat transfer capacity of the cooling tower is controlled in part load operation by adjusting the fan speed.

The inside of the cooling tower is filled with packing material such as porous plastic sheets to maximize the contact surface area between the air and the water. The water is cooled since it is used to provide the latent heat of evaporation of the water film and droplets. Make up water is needed to replace the evaporative water loss. The flow rate of water is maintained high enough so that the make up water flow rate is a very small proportion of the overall water flow rate.

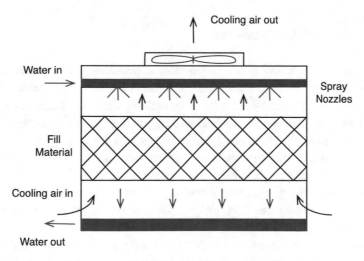

Fig. 7.7 Schematic of induced-draft cooling tower

As discussed previously, with direct evaporative cooling, the lowest temperature that the water can reach is the wet bulb temperature of the incoming air. The 1% wet bulb design temperature for a given location is usually chosen for the design entering air condition.

7.8   Compressors

The types of compressors used in refrigeration and air conditioning applications can be grouped into five categories: reciprocating, centrifugal, rotary, scroll, and screw. The type of compressor chosen depends on the specific application, capacity and efficiency required. Rotary and scroll compressors are used in smaller capacity residential and vehicular air conditioning and refrigeration applications. Reciprocating, centrifugal, and screw compressors are used in larger capacity commercial air conditioning and refrigeration.

Compressor capacity maps are provided by manufacturers for compressor selection to meet a given cooling load. There is capacity overlap between the compressor categories. Rotary compressors are used in cooling systems requiring between 1 and 5 kW, reciprocating compressors between 1 and 500 kW, scroll compressors between 5 and 50 kW, screw compressors between 50 and 500 kW, and centrifugal compressors between 200 and 10,000 kW. Small capacity compressors in residential applications are bundled with an electric motor in a common package called a hermetic compressor. The refrigerant is used to cool the electric motor so the compression process is not adiabatic.

7.8.1   Reciprocating Compressors

The reciprocating compressor uses a piston-cylinder geometry, and is the most commonly used compressor type. As shown in Fig. 7.8, it consists of a crankshaft connected to a piston by a connecting rod. Driven by the rotating crankshaft, the piston oscillates up and down in the cylinder with stroke s and bore b. Depending on the capacity required, reciprocating compressors can have more than one cylinder, driven by a single crankshaft.

Valves are used to control the flow of refrigerant vapor into and out of the cylinder. The valves are of reed or poppet type design. The reed valves operate by differential pressure, so the inlet valve does not open until the pressure in the cylinder drops below the pressure in the inlet line from the evaporator. Piston rings are used to seal the piston and maintain the compression pressure in the cylinder.

The compression process is shown on the $P - V$ diagram of Fig. 7.9. From state 1–2 the refrigerant enters the cylinder through the inlet valve during the inlet or suction stroke. At bottom dead center, the piston speed goes to zero, so the inlet flow and pressure drop across the inlet valve also decrease. As the piston moves upward, the cylinder pressure increases, the inlet valve closes, and the refrigerant is compressed during the compression stroke from

Fig. 7.8 Piston-cylinder geometry

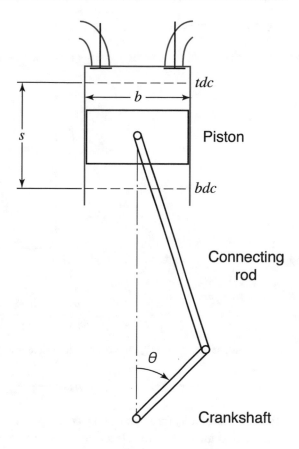

2 to 3. Near top dead center, the outlet valve opens and the compressed refrigerant flow into the outlet manifold.

The clearance volume at state 4 will contain residual refrigerant which will mix with the inlet gas when the inlet valve opens. As the piston moves away from top dead center, no new gas will be drawn into the cylinder until the cylinder pressure drops to the suction pressure at state 1.

The displacement volume V_d is defined as the difference between the maximum and the minimum volumes,

$$V_d = V_2 - V_4 = \frac{\pi}{4}b^2 s \tag{7.19}$$

The clearance factor c is the clearance volume relative to the displacement volume, and typically ranges from 0.02 to 0.08.

$$c = \frac{V_4}{V_d} \tag{7.20}$$

The volumetric efficiency η_v is defined as the actual mass m_i of vapor compressed relative to the mass that would fill the displacement volume at the inlet density ρ_i.

Fig. 7.9 The compression process

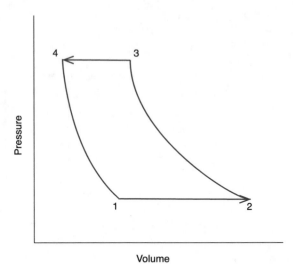

$$\eta_v = \frac{m_i}{\rho_i\, V_d} = \frac{\bar{\rho}}{\rho_i}\, \frac{(V_2 - V_1)}{V_d} \tag{7.21}$$

For a compressor speed N, the volumetric efficiency is

$$\eta_v = \frac{\dot{m}}{\rho_i\, V_d\, N} \tag{7.22}$$

Typical values of η_v for reciprocating compressors range from 0.75 to 0.9. The volumetric efficiency is affected by the clearance factor, the pressure drop across the inlet valve, heat transfer from the cylinder walls, and gas leakage past the piston rings, which will decrease the density of the inlet charge during during the intake stroke from ρ_i to an average value $\bar{\rho}$. One method of control for reciprocating compressors is to vary the clearance factor by opening/closing clearance valve pockets which will change the volumetric efficiency and thus the refrigerant flowrate.

We can do a simple calculation to estimate the effect of the above parameters on the volumetric efficiency by assuming a polytropic expansion with exponent k of the residual gas from state 4 to state 1:

$$V_1 = V_4 \left(\frac{P_4}{P_1}\right)^k \tag{7.23}$$

The term $V_2 - V_1$ is

$$
\begin{aligned}
V_2 - V_1 &= (V_2 - V_4) - (V_1 - V_4) \\
&= (V_2 - V_4) - V_4\left((P_4/P_1)^{1/k} - 1\right) \\
&= V_d - c\, V_d\left((P_4/P_1)^{1/k} - 1\right)
\end{aligned}
\tag{7.24}
$$

Thus, upon substitution of Eq. (7.21), the volumetric efficiency can be expressed as

$$\eta_v = \left[1 + c\,(1 - (P_4/P_1)^{1/k} \right] \frac{\bar{\rho}}{\rho_i} \tag{7.25}$$

Example 7.4 Reciprocating Compressor Volumetric Efficiency
A reciprocating compressor using R-134a as the refrigerant has a clearance factor $c = 0.05$. The discharge pressure is 770 kPa. The evaporator operates at -7.0 C (266 K) and 250 kPa. The average temperature of the gas in the cylinder during the intake stroke is $4\,°C$ (277 K), and there is a pressure drop of 10 kPa across the inlet valve. Assume k for the residual gas expansion is 1.14. What is the volumetric efficiency η_v for the above conditions?

Solution
For R-134a, the gas constant R is

$$R = \frac{R_u}{M} = \frac{8.314}{102.3} = 0.0813\,\text{kJ/kg K}$$

The inlet refrigerant density is

$$\rho_i = \frac{P_i}{R\,T_i} = \frac{100}{(0.813)(266)} = 4.62\,\text{kg/m}^3$$

The average refrigerant density in the cylinder is

$$\bar{\rho} = \frac{P_i - \Delta P}{R\,\bar{T}} = \frac{100 - 10}{(0.0813)(277)} = 4.00\,\text{kg/m}^3$$

The volumetric efficiency is

$$\eta_v = \left[1 + c\,(1 - (P_4/P_1)^{1/k} \right] \frac{\bar{\rho}}{\rho_i}$$

$$= \left[1 + 0.05\,(1 - (770/250)^{1/1.14}) \right] \frac{4.00}{4.62}$$

$$= 0.79$$

7.8.2 Centrifugal Compressors

The centrifugal compressor is a turbomachine with a continuously flowing working fluid. It uses a rotating impeller to add tangential velocity to the working fluid which is converted to a head or pressure rise in a scroll or volute through an increase in the flow cross sectional area. Since the flow through the compressor is continuous, centrifugal compressors have a greater volumetric capacity than positive displacement compressors. The capacity of a

Fig. 7.10 Schematic of
centrifugal compressor

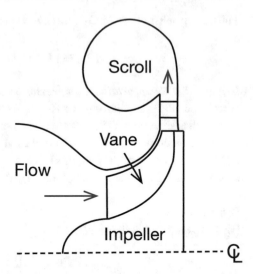

centrifugal compressor is very sensitive to the evaporator and condenser temperatures of a
vapor compression system.

Centrifugal compressors are sized to meet 200–10,000 kW of cooling capacity. The larger
centrifugal compressor systems are built up as package units comprising the condenser,
evaporator, and compressor into a single unit. Typically two or more impeller stages are
used in HVAC applications, since the diameter of the impeller would be too large to operate
at the tip velocity required for a single stage. Depending on the pressure ratio required, the
compression can be up to eight stages. The impeller rotational velocity is usually in the
3500–7600 rpm range.

A schematic diagram of the side of a single stage centrifugal compressor is given in
Fig. 7.10, and a schematic of a section of the impeller is is given in Fig. 7.11. As shown in
Fig. 7.10, the refrigerant gas enters the impeller axially, and exits radially.

Using the control volume momentum and energy equations, we can develop relations
for the power consumption and pressure rise in a centrifugal compressor as a function of
the flow rate and impeller geometry. Inlet and exit velocity diagrams across the compressor
impeller are given in Fig. 7.12. The absolute velocity is given by V, the velocity of the flow
relative to the impeller is given by W, and the impeller velocity is $U = \omega r$.

For the impeller blade geometry, the inlet blade angle is β_1, and the exit blade angle is β_2.
For the design load condition, the tangential component of the inlet velocity is V_{1t} is usually
assumed to be zero, so $V_{1n} = V_1$. For part load conditions, inlet vanes in the compressor
housing upstream of the impeller are used to impart a swirl to the inlet flow which will
reduce the angular momentum added to the flow. The inlet portion of the blades are forward
curved, with $\beta_1 < 90°$, so the relative velocity of the incoming flow is tangent to the vanes,
as shown in Fig. 7.12.

From the inlet velocity triangle,

Fig. 7.11 Schematic of
impeller section

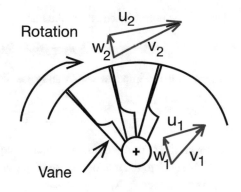

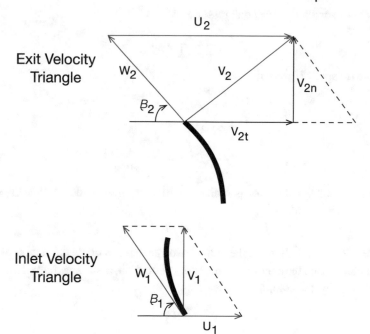

Fig. 7.12 Inlet and exit velocity diagrams

$$\tan \beta_1 = \frac{V_1}{\omega \bar{r}_1} \tag{7.26}$$

and from the outlet velocity triangle,

$$\tan \beta_2 = \frac{V_{2n}}{\omega r_2 - V_{2t}} \tag{7.27}$$

Using the continuity equation, the mass flowrate through the impeller exit is

$$\dot{m} = \rho_2 A_2 V_{2n} = \rho_2 2\pi r_2 b_2 V_{2n} \tag{7.28}$$

where b_2 is the impeller width, and r_2 is the impeller tip radius. Solving for V_{2n},

$$V_{2n} = \frac{\dot{m}}{\rho_2 2\pi r_2 b_2} \tag{7.29}$$

From the control volume momentum equation, the torque T from the impeller onto the fluid is equal to the rate of change of angular momentum of the fluid. If the incoming swirl V_{1t} is zero, then

$$T = \dot{m}(r_2 V_{2t} - r_1 V_{1t}) = \dot{m}(r_2 V_{2t}) \tag{7.30}$$

The power $-\dot{W}$ supplied to the compressor is

$$-\dot{W} = T\omega \tag{7.31}$$

Since the tip speed $U_2 = \omega r_2$, then

$$\begin{aligned} -\dot{W} &= \dot{m} r_2 V_{2t}\,\omega \\ &= \dot{m}\, U_2 V_{2t} \\ &= \dot{m}\, U_2^2 \left(1 - \frac{V_{2n}}{U_2 \tan\beta_2}\right) \end{aligned} \tag{7.32}$$

For the case where the exit flow is radial, i.e., the impeller blade angle is $\beta_2 = 90°$, the compressor power is simply

$$-\dot{W} = \dot{m} U_2^2 \tag{7.33}$$

We now develop an equation for the compressor pressure ratio as a function of the above flow parameters. From the control volume energy equation, assuming the exit scroll velocity $\simeq$ the inlet velocity, the compressor power is

$$-\dot{W} = \dot{m}(h_2 - h_1) \tag{7.34}$$

Equating Eqs. (7.32) and (7.34) for the compressor power yield an equation for the exit tangential velocity V_{2t},

$$V_{2t} = \frac{h_2 - h_1}{U_2} \tag{7.35}$$

For an isentropic compression from the inlet at station 1 to the exit at station 2, the temperature ratio is

$$\frac{T_{2s}}{T_1} = \left(\frac{P_2}{P_1}\right)^{(\gamma-1)/\gamma} \tag{7.36}$$

and the corresponding isentropic compression efficiency is

$$\eta = \frac{h_{2s} - h_1}{h_2 - h_1} \tag{7.37}$$

Upon substitution of Eqs. (7.36) and (7.37) the compressor power $-\dot{W}$ can be expressed as

$$\begin{aligned}
-\dot{W} &= \dot{m}\,(h_{2s} - h_1)/\eta \\
&= \dot{m}\,c_p(T_{2s} - T_1)/\eta \\
&= \dot{m}\,c_p\,T_1 \left[\left(\frac{P_2}{P_1} \right)^{(\gamma-1)/\gamma} - 1 \right]/\eta
\end{aligned} \tag{7.38}$$

Equating Eqs. (7.32) and (7.38) results in Eq. (7.39) for the compressor pressure ratio P_2/P_1:

$$\frac{P_2}{P_1} = \left[\frac{\eta\,U_2^2\,(1 - \frac{V_{2n}}{U_2\,\tan\beta_2})}{c_p T_1} + 1 \right]^{\gamma/(\gamma-1)} \tag{7.39}$$

This is the maximum pressure ratio that can be produced for a given tip speed and efficiency.

A representative pressure ratio-flowrate plot for a centrifugal compressor is given in Fig. 7.13. Mass flow is given on the x-axis and pressure ratio on the y-axis. Contour lines of constant adiabatic efficiency and constant speed are also plotted. Note that as the flowrate increases at a given compressor speed, the pressure ratio decreases.

Centrifugal compressors have surge and choking performance limits. The surge limit on the left side of the compressor map represents a boundary between stable and unstable operating points. For stable operation compressors operate to the right of the surge line, with

Fig. 7.13 Representative centrifugal compressor map

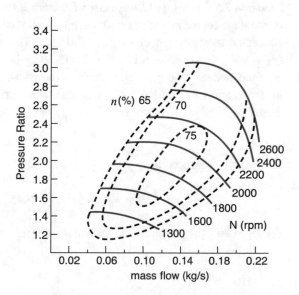

a negative slope to the constant speed lines. Surge is a self-sustaining flow oscillation. When the mass flow rate is reduced at constant pressure ratio, a point arises where somewhere within the internal boundary layers on the compressor blades a flow reversal occurs. If the flow rate is further reduced, then a complete reversal occurs which relieves the adverse pressure gradient. That relief means a flow reversal is no longer needed and the flow then begins to return to its initial condition. When the initial condition is reached, the process will repeat itself, creating surge. This feature makes the centrifugal compressor very sensitive to the condenser pressure. Hot gas bypass is used to extend the surge limit.

On the right side of the dynamic compressor map is a zone where efficiencies fall rapidly with increasing mass flow rate. The gas speeds are quite high in this zone and the attendant fluid friction losses are increasing with the square of the gas speed. In this region there is also the choke limit which occurs at a slightly different value of $\dot{m}/\dot{m}_{cr}$ for each tip speed. Choking occurs when at some point within the compressor the flow reaches the speed of sound. It occurs at values of $\dot{m}/\dot{m}_{cr}$ less than 1 because $\dot{m}_{cr}$ is based on the compressor wheel diameter D rather than on the cross sectional area where choking is occurring. The value of $\dot{m}/\dot{m}_{cr}$ at choking varies with tip speed because the location within the compressor at which choking occurs depends on the structure of the internal boundary layers.

The centrifugal compressor performance curves can also be used for system control. For example, in the case of a compressor with constant condenser pressure P_2, as the external load decreases, the evaporator pressure P_1 will decrease, the pressure ratio will correspondingly increase, and the compressor will operate at a lower flow rate, following the constant speed line. When it is necessary to decrease the flow rate at a constant evaporator temperature T_1, a similar result can be produced by increasing the condenser temperature T_2, which is accomplished by reducing the condenser water flow rate.

Example 7.5 Centrifugal Compressor Pressure Ratio

A two stage centrifugal compressor has an impeller with a radius of 0.45 m which is directly connected to a motor operating at 3350 rpm. The exit flow from the impeller is radial ($\beta_2 = 90°$). The compressor isentropic efficiency is 0.80. Refrigerant R-134a is the working fluid, and the saturation temperature in the evaporator is 0 °C. What is the pressure ratio in the compressor, and the corresponding condenser saturation temperature? Assume the thermophysical properties of R-134a are $c_p = 852$ J/kg, and $\gamma = 1.106$.

Solution
The impeller tip speed is

$$U_2 = \omega r_2 = 3350 \frac{2\pi}{60} 0.45 = 157.9 \, \text{m/s}$$

For $\beta_2 = 90°$, the pressure ratio for one stage from Eq. (7.39) is

$$\frac{P_2}{P_1} = \left[\frac{\eta\, U_2^2}{c_p T_1} + 1\right]^{\gamma/(\gamma-1)}$$

$$= \left[\frac{0.80\,(157.9)^2}{(852)\,(273)} + 1\right]^{1.106/(1.106-1)}$$

$$= 2.36$$

For two stages, the pressure ratio is therefore $(2.36)\,(2.36) = 5.57$. From R-134a tables, the saturation vapor pressure at $0\,^\circ$C is 244.5 kPa, so

$$P_2 = 5.57\, P_1 = (5.57)(244.5) = 1361 \text{ kPa}$$

The saturation temperature at that pressure is $T_2 = 51.3\,^\circ$C.

7.8.3 Compressor Lubrication

Since compressors are mechanical devices with metal parts that slide or rotate, they require lubrication to reduce piston and bearing friction, and increase refrigerant sealing. Lubrication oils are paraffin, i.e., mineral oil based.

During the compression stroke of a reciprocating compressor, some of the lubricating oil on the cylinder wall will be entrained as droplets and will pass out with the discharge gas. An oil separator is frequently used in the compressor discharge line to reduce the amount of entrained oil. The separation process is accomplished by droplet impingement on a flat plate or centrifugal rotation, typically with 95–98% efficiency.

The remaining oil will pass through the condenser, and the expansion valve, and enter the evaporator. In the evaporator, the refrigerant will vaporize, and the oil will remain liquid. The amount of oil accumulation depends on the miscibility of the oil and refrigerant. Lubrication oil is fully miscible with R-134a, and non-miscible with R-717 (Ammonia). With ammonia as the working fluid, the accumulated lubrication oil will sink to the bottom of the condenser and evaporator, and can be drained directly.

7.9 Expansion Valves

Expansion devices are used in vapor compression systems to create the pressure drop between the high and low pressure sides of the system, and also control the refrigerant flow rate. The expansion valve controls the flow of the refrigerant from the high pressure condenser to the low pressure evaporator. The type of expansion valve chosen depends on the evaporator operation. Major types are a fixed area restrictor, constant pressure expansion valve, thermostatic expansion valve, and electric expansion valve.

Fig. 7.14 Cross-section of
thermostatic valve

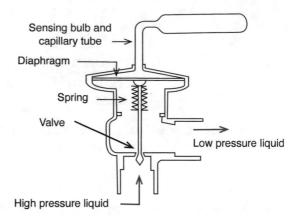

Flooded evaporators use a solenoid valve for the pressure reduction coupled to a float valve for on-off control of the liquid level in the evaporator. When the liquid level in the evaporator reaches the set point, the float valve closes, shutting of the flow of refrigerant to the evaporator. As the liquid level decreases, the float valve opens, allowing refrigerant to flow into the evaporator.

Dry expansion evaporators use thermostatic expansion valves, designed to maintain a constant amount of superheat in the vapor at the evaporator outlet. Since it is important to avoid liquid refrigerant entering the compressor, the refrigerant leaving the evaporator is superheated by about 5 °C. As indicated in Fig. 7.14, the valve contains a moveable diaphragm attached to a spring and valve, sensitive to the pressure taps above and below the diaphragm. A temperature and pressure sensor bulb is located at the outlet of the evaporator which is connected above the diaphragm, so the pressure above the diaphragm is equal to the superheated pressure at the evaporator outlet. The pressure below the diaphragm is the pressure at the inlet of the evaporator, i.e., the saturation pressure of the refrigerant.

The operation of the thermostatic expansion valve is as follows. The diaphragm moves in response to the difference in pressures across the diaphragm. An adjustable spring balances this pressure differential so that if the pressure differential changes, the spring will open or close the valve accordingly. For example, as the temperature and thus the pressure in the bulb increases due to an increase in the cooling load, the pressure above the diaphragm increases, the valve will open further, and increase the flow of refrigerant to the evaporator. An increase in refrigerant flowrate will lower the evaporator exit temperature back to the set point.

Small capacity systems such as domestic refrigerators use fixed area capillary tubes. which are small diameter tubes used to produce the required pressure drop. They are passive devices sized for the required pressure drop for a single operating point, and have a limited ability to regulate refrigerant flow in response to changes in cooling load. The pressure drop in the capillary tube can be predicted accurately for a known fluid viscosity. As the pressure of the fluid along the capillary tube decreases, some of the liquid refrigerant will vaporize, resulting in a two phase liquid flow in the tube. A fixed orifice can also be used to produce the needed pressure drop. Additional system control is achieved by on-off motor control.

Since the expansion valve creates an energy loss in the cooling cycle, there have been attempts to use turbines instead of valves to extract work from the pressure drop, similar to a gas turbine or Brayton cycle. However, while technically feasible, there is additional complexity and cost which makes such approaches not commercially viable.

7.10   Further Reading

Heat exchanger design is covered extensively in Shah (2002). Additional information about boiling and condensation processes is given in Collier (1996) and Carey (2020). Cooling towers are discussed further in Hill (2013). The design and analysis of centrifugal compressors is covered in Braembussche (2019).

References

ASHRAE (2017) ASHRAE Handbook - Fundamentals, Chapter 30 - Thermophysical Properties of Refrigerants. ASHRAE, Atlanta, Georgia

Braembussche R (2019) Design and Analysis of Centrifugal Compressors. John Wiley, New York

Carey V (2020) Liquid-Vapor Phase Change Phenomena. CRC Press, Boca Raton, Florida

Collier J, Thome J (1996) Convective Boiling and Condensation. Oxford Univ Press, London

Dobson M, Chato J (1998) Condensation in smooth horizontal tubes. ASME J Heat Transfer 120:193–213

Hill G (2013) Cooling Towers: Principles and Practice. Elsevier, Oxford, England

Rohsenow W, Hartnett J, and Cho Y (1998) Handbook of Heat Transfer. 1570-1571, McGraw-Hill, New York

Shah R, Sekulic D (2002) Fundamentals of Heat Exchanger Design. John Wiley, New York

Physical Constants and Conversion Factors

A

A.1   Air and Water Properties

See Table A.1.

Table A.1 Properties of Air and Water at $20\,^{\circ}$ C (293 K) and 1 atm

Property	Units	Air	Water
ρ	(kg/m^3)	1.20	998.2
c_p	(kJ/kg-K)	1.004	4.183
k	(W/m-K)	0.0251	0.586
Pr	(-)	0.707	6.99
μ	(Pa-s)	18.25×10^{-6}	1.0×10^{-3}
ν	(m^2/s)	15.89×10^{-6}	1.0×10^{-6}

© The Editor(s) (if applicable) and The Author(s), under exclusive license to Springer
Nature Switzerland AG 2023
A. T. Kirkpatrick, *Introduction to Refrigeration and Air Conditioning Systems*,
Synthesis Lectures on Mechanical Engineering,
https://doi.org/10.1007/978-3-031-16776-8

A.2   Physical Constants

See Table A.2.

Table A.2 Physical Constants

Universal Gas Constant	R_u	$= 8.314$ kJ/kmol-K
		$= 8.314$ Pa m^3/mol-K
		$= 8.314 \times 10^{-2}$ m^3 bar/kmol-K
		$= 8.205 \times 10^{-2}$ m^3 atm/kmol-K
		$= 1545$ ft lb$_f$/lbmole-$^\circ$ R
		$= 1.986$ Btu/lbmole-$^\circ$ R
Avogadro's Number	N	$= 6.024 \times 10^{23}$ molecules/mol
Planck's Constant	h	$= 6.625 \times 10^{-34}$ J-s/molecule
Boltzmann's Constant	k	$= 1.380 \times 10^{-23}$ J/K-molecule
Speed of Light in Vacuum	c_o	$= 2.998 \times 10^{8}$ m/s
Stefan-Boltzmann Constant	σ	$= 5.670 \times 10^{-8}$ W/m^2-K^4
		$= 0.1714 \times 10^{-8}$ Btu/h ft^2-$^\circ$ R^4
Gravitational Acceleration	g	$= 9.807$ m/s^2
Standard Atmospheric Pressure	P	$= 101,325$ N/m^2
		$= 101.325$ kPa

A.3   Conversion Factors

See Table A.3.

Table A.3 Unit Conversion Factors

Area	1 m^2	$= 1550.0$ in^2
		$= 10.764$ ft^2
Energy	1 J	$= 9.4787 \times 10^{-4}$ Btu
		$= 0.239$ calories
Energy/Mass	1kJ/kg	$= 0.4303$ Btu/lb$_m$
Force	1 N	$= 0.22481$ lb$_f$
Heat transfer rate	1W	$= 3.4123$ Btu/h
Heat flux	1 W/m^2	$= 0.3171$ Btu/h-ft^2

(continued)

Table A.3 (continued)

Heat transfer coefficient	$1 \text{ W/m}^2\text{-K}$	$= 0.17612 \text{ Btu/h-ft}^2 \, ^\circ\text{F}$
Thermal Diffusivity	$1 \text{ m}^2/\text{s}$	$= 3.875 \times 10^4 \text{ ft}^2/\text{h}$
Length	1 m	$= 39.370 \text{ in.}$
		$= 3.2808 \text{ ft.}$
	1 km	$= 0.62137 \text{ mile}$
Mass	1 kg	$= 2.2046 \text{ lb}_m$
Mass density	1 kg/m^3	$= 0.062428 \text{ lb}_m/\text{ft}^3$
Mass flow rate	1 kg/s	$= 7936.6 \text{ lb}_m/\text{h}$
Mass transfer coefficient	1 m/s	$= 1.1811 \times 10^4 \text{ ft/h}$
Power	1 kW	$= 1.341 \text{ hp}$
Pressure and stress	$1 \text{ Pa} \ (1 \text{ N/m}^2)$	$= 0.020886 \text{ lb}_f/\text{ft}^2$
	$1.01325 \times 10^5 \text{ N/m}^2$	$= 1 \text{ standard atmosphere}$
		$= 760 \text{ mmHg}$
	$1 \times 10^5 \text{ N/m}^2$	$= 1 \text{ bar}$
		$= 750.06 \text{ mmHg}$
Rotational speed	1 rev/min	$= 0.10472 \text{ rad/s}$
Specific heat	1 J/kg-K	$= 2.3886 \times 10^{-4} \text{ Btu/lb}_m\text{-}^\circ\text{F}$
Temperature	K	$= \,^\circ\text{C} + 273.15$
		$= (5/9) \, (^\circ\text{R} + 459.67)$
Temperature difference	1 K	$= 1 \,^\circ\text{C}$
		$= (9/5) \,^\circ\text{R} = (9/5) \,^\circ\text{F}$
Thermal conductivity	1 W/m-K	$= 0.57782 \text{ Btu/h ft-}^\circ\text{F}$
Thermal resistance	1 K/W	$= 0.52750 \,^\circ\text{F/h-Btu}$
Torque	1 Nm	$= 0.73756 \text{ lb}_f \text{ ft}$
Viscosity (dynamic)	$1 \text{ kg/m-s}^2 \ (\text{Pa-s})$	$= 2419.1 \text{ lb}_m/\text{ft-h}$
Viscosity (kinematic)	$1 \text{ m}^2/\text{s}$	$= 3.875 \times 10^4 \text{ ft}^2/\text{h}$
Volume	$1 \text{ m}^3 \ (10^3 \text{ L})$	$= 6.1023 \times 10^4 \text{ in.}^3$
		$= 35.314 \text{ ft}^3$
		$= 264.17 \text{ gal}$
Volume flow rate	$1 \text{ m}^3/\text{s}$	$= 2.1189 \times 10^3 \text{ ft}^3/\text{min}$
		$= 1.5850 \times 10^4 \text{ gal/min}$

Printed in the United States
by Baker & Taylor Publisher Services